The New Race for Space

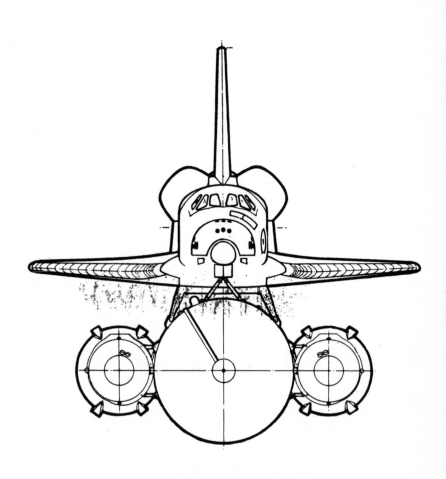

The New Race for Space

The U.S. and Russia Leap to the Challenge for Unlimited Rewards

James E. Oberg

Foreword by Ben Bova

Stackpole Books

The opinions and evaluations expressed in this book are those of the author and do not necessarily reflect those of McDonnell Douglas, NASA, or any other organization.

Printed in the U.S.A.

The author and publisher of this book are grateful to: OMNI magazine for permission to use material that appeared in a somewhat different form in the following OMNI articles:
"Beyond Sputnik's Booster" © 1982 by Omni Publications International, Ltd.·
"Hitch Up to a Red Star" © 1982 by Omni Publications International, Ltd.
"Attack of Cosmonauts" © 1984 by Omni Publications International, Ltd.
"Mars Manifesto" © 1981 by Omni Publications International, Ltd.
"Return to the Moon" © by Omni Publications International, Ltd.
"Space Rope Tricks" © by Omni Publications International, Ltd.
"Red Wings in Orbit" © by James E. Oberg.
"Spaceships of the Future" © by James E. Oberg.
"Geostationary Man" © 1984 by James E. Oberg.

SCIENCE DIGEST for permission to use material that first appeared in a somewhat different form in the following SCIENCE DIGEST articles:
"Comrades in Orbit" © 1983 by The Hearst Corporation.
"Souping Up the Space Shuttle" © by The Hearst Corporation.

POPULAR MECHANICS for permission to use material that first appeared in a somewhat different form in the following POPULAR MECHANICS article:
"Mining the Asteroids" © by The Hearst Corporation.

Library of Congress Cataloging in Publication Data

Oberg, James E., 1944–
    The new race for space.

    1. Astronautics—United States.   2. Astronautics—
Soviet Union.   I. Title.
TL789.8.U5024   1984   629.4'0947   84-2576
ISBN 0-8117-2177-9

To Cooky, who made
it possible—and
worthwhile.

Contents

Foreword

My favorite psychologist tells me that every person can be identified by at least three different images: the image that the person has of himself; the image that the person wants others to see; and the image that others actually perceive. In other words: who we think we are, who we want others to think we are, and who others really think we are.

This multiplicity of images applies to other things besides people. In particular, it applies to the great, complex interaction of human beings and machines that we call "the space program." In effect, there are three space programs: the space program as it really is; the space program as the public (through the eyes of the news media) perceives it; and the space program as enthusiasts like Jim Oberg and I wish it could be.

This book has a great deal to say about all three. And I can think of no one better qualified to say it than Jim Oberg, who for years has worked in the U.S. military and civilian space programs, and has closely followed the performance of the Russian space program. Jim knows what actually exists. And as this book makes clear, the reality of both the American and Soviet space programs is rather different from the images projected by the news media—and different, still, from what we could be achieving if we put our whole hearts and minds to the task.

A word about the news media. They are the eyes and ears of the public; their reports on space activities determine to a major extent what the public knows about the subject and, in the western democracies, how much taxpayer support there will be for space efforts.

For years it has been obvious to me that the media covers the space program pretty much the way they cover the Ringling Brothers–Barnum & Bailey Circus. Think about it. Every year, when the circus comes to town, there is a flurry of stories in the local newspapers and on TV. The circus makes "good copy": elephants and clowns and lion tamers and trapeze artistes [*sic*]. For the week or so that the circus is around, the media covers the story.

And then nothing else is heard about the circus until next year when it comes to town again—unless there is a disaster somewhere: a fire in the Big Tent or an elephant going berserk. If no disaster occurs, the circus disappears from the media's attention as completely as if it had vanished from the face of the Earth.

That is exactly how the media covers the space program. When there is a launch scheduled, especially the launch of a new type of rocket, the media swarms in and almost overkills the story with coverage. The launch of the first few space shuttle missions, for example, were covered and analyzed and favored with "instant replays" almost as heavily as the Super Bowl. But between launches there was practically no coverage whatsoever of the space program—unless a disaster struck: troubles with a new rocket engine; problems with heat shield tiles.

The launches are spectacular and photogenic. But the *real* work of the space program goes on between the launches. The budgetary battles that have broken brave men's spirits. The missed opportunities. The decision to kill the Apollo lunar program precisely at the time when the United States had achieved such a commanding presence in space that everything between here and the Moon was *mare nostrum*. The cutbacks that forced priceless trained talent to quit the program and seek employment elsewhere.

And the media goes on, using circus-story attitudes in their decisions about what is "news" in the space program. The shuttle is so successful that launches are coming almost every month. Who wants to see another launch on the six o'clock news, or on the front page of the morning paper? The crowds at Cape Canaveral are down from a million per shot to only half a million; obviously public interest in the shuttle is waning.

When the President announces that our next goal in space will be to build a permanent manned station in orbit, the media reports that it will cost almost $10 billion but fails to report how much such a station will add to the U.S. Gross National Product. The reporters who use transistorized tape recorders and miniaturized videotape cameras and desktop microcomputer word processors fail to understand that their everyday tools are products of space technology.

And where their coverage of the American space program is lax, their coverage of the space programs of other nations—particularly the Soviet Union's—is virtually nonexistent.

The New Race for Space is an important book because it shows the space programs of both the United States and the USSR as they actually are—and as they may one day be. Since the final decisions on the size, scope, and usefulness of the American space program rests in your hands, as a voter and taxpayer, this book is required reading for every good citizen. And it is also a fascinating and thoroughly enjoyable look at the marvelous achievements of yesterday and the even more marvelous possibilities of tomorrow. Jim Oberg has done us all a great favor.

Ben Bova
President, National Space Institute

Acknowledgments

Portions of the material in this book appeared, in preliminary form, in numerous magazines around the world. Making it cogent and readable was aided by my editors, including in particular Gurney Williams of *OMNI*, Scott DeGarmo of *Science Digest*, Mike Kenward of *New Scientist* (London), Wayne Matson of *Aviation/Space*, Shigeru Komori of *Quark* (Tokyo), Robert Cowen and Pete Spotts of the *Christian Science Monitor*, and Terry Dickinson of the late, lamented *Star and Sky*. Substantial new material has been added, and old material has been updated extensively. In book form, my Stackpole editor, Peggy Senko, performed a marvelous job of tying the disparate elements together where I frequently overlooked problems.

In the area of Soviet research, which occupies almost half the book, I continue to be grateful for support from the loose-knit "Kettering Group," including Geoffrey Perry, Mark Severance, Dick Flagg, Sven Grahn, along with free-lance Soviet sleuths such as Nicholas Johnson, Charles Patrick Vick, Rex Hall, Saunders Kramer, and government employees Ed Zvetina, Nick Timacheff, Marcia Smith, Joe Rowe, and Joe Bruman.

xiii

For NASA material I am indebted for support from Terry White, John Lawrence, Mike Gentry, Lisa Vazquez, Lee Saegesser, Hu Harris, and Ed Harrison.

As an aid to the relative ease with which this book was composed, hats off to the folks who designed and built the Osborne-1 and Wordstar!!

1

The Invisible Space Race

Recent space activities by the United States and the Soviet Union have shown a markedly divergent philosophical approach. The American space shuttle program is providing the promised easy access to and from orbit for quick trips; the Soviet Salyut space station is providing the long-anticipated permanent human presence in orbit.

To all appearances, neither nation is pursuing any major goals shared by the other. In the 1960s, common goals (e.g., the Moon) allowed observers to make easy judgments about relative standings in the "space race." But by that metaphor, today's Soviet and American space programs are not even racing; they are on entirely different race tracks.

But if there is no race, why are both the United States and the Soviet Union still running so hard?

The visible differences in the current space strategies of the United States and the USSR are due to less apparent but still real differences in the services which both countries seek to obtain from space. These services, in turn, reflect the technological capabilities and weaknesses of each nation. It is this difference in capabilities which results in the

1

apparent differences in strategies—even though both nations are now
seeking to gain many of the same practical advantages from space
activities.

The USSR's goal of a permanent manned facility in orbit is ne-
cessitated by the limited lifetime of unmanned Soviet spacecraft. The
USSR must launch more than one hundred satellites every year because
their space vehicles break down within a short period. No more than
eighty Soviet spacecraft are really "alive" at any time.

For example, the failure of Soviet deep-space probes to endure
long interplanetary voyages is the key reason that the USSR has had
success only with one planet, Venus, to which the travel time is only
four months. And this is one reason for considerable skepticism among
some observers about the success of the Soviet Halley's comet missions
which will be en route for fifteen months, an unprecedented duration.

Under these circumstances, the Soviets can most easily increase
the utility of their space activity by extending the lifetimes of space
vehicles rather than by lowering the cost of individual launchings.
Fundamental limitations in the manufacturing quality of Soviet electronics equipment have frustrated attempts to lengthen the operation
of automatic equipment. The answer is to build a manned platform
where cosmonauts can tend equipment aboard the space station or in
orbits nearby—including resupply, preventive maintenance, repair, and
retrieval of experimental results as needed.

For American space activities, almost the opposite is the problem.
Individual space vehicles last for many, many years. Fewer launchings
are thus required (perhaps a dozen, or twenty, per year), and the cost
of each additional launching is consequently higher in relative terms
than it is for the USSR. Since replacement satellites may wait a long
time "in the pipeline" before they are launched, the introduction of
technological improvements can become very expensive.

The answer NASA has chosen is to provide a launch system flexible enough to carry almost any type of spacecraft or equipment, with
quick response time to changing requirements, and with the presence
of people *not* for repair of breakdowns but for making sure that the
satellite is deployed correctly and in "good health." Lengthy tours of
duty in space are not required for these tasks.

Official commentary in both major spacefaring nations has stressed
that this is "the decade of space exploitation," as compared to earlier
periods of pure "space exploration." This reflects a common desire to
gain more practical benefits from space activities.

In the past, services such as communications and weather forecasting have gained enormously from activities of space vehicles; up

until now, space military activities have contributed to each nation's security and to diplomatic stability. But the main developments in coming years will be space industrialization and space-based earth observation. There are conditions which are available for free in space but which industrial nations pay billions of dollars each year to duplicate; there are features of the Earth's surface, oceans, and atmosphere which only become visible when observers travel some distance away from them. The mastery of these technologies has measurable economic value, and the value is very large.

Whichever nation comes out ahead in this competition will be the winner of the current round of the space race—a race no longer for glory, or curiosity, but for wealth and power. Each nation brings its own capabilities and shortcomings into this competition. But the space programs of both the United States and the USSR have begun to allow other nations to participate actively and to share in the wealth and power to be won in space. This factor, the fading of the bilateral nature of the space race into a multilateral effort, is another major theme of space activity in the 1980s and beyond.

THE SOVIET ADVANTAGE

The immediately obvious Soviet advantage is the existence of a prototype permanent manned space station: the Salyut complex. Cosmonauts can occupy this facility for months on end and be replaced by fresh space teams before returning to Earth. Recently, chief cosmonaut Gen. Vladimir Shatalov declared, "We are very close to round-the-clock and year-round operations of space stations"—and that moment may already have arrived by the time these words are read.

Aboard the Salyut, cosmonauts have engaged in a broad program of industrial experiments. Small electrical furnaces have been used to create special crystals, metal alloys, and glass products. Another small unit, the Tavriya, has been used for separation of biologically active compounds, thus promising increased purity and availability of rare pharmaceutical products (see Chapter 7).

From on board the Salyut, cosmonauts have observed their native planet with a powerful array of instruments. The MKF-6M multi-spectral camera (manufactured by the Karl Zeiss Jena plant in East Germany) has allowed the survey of croplands and the detection of unknown geological features which hint at rich ore and oil deposits (see Chapter 7). Mapping cameras have done more topographic surveillance in ten minutes than a camera-carrying aircraft could do in a year. Naked-eye observations of the subtle color variations of the ocean

have allowed cosmonauts to pinpoint potentially plankton-rich regions
and to direct fishing fleets to them. (In 1982 alone, the Soviet Ministry
of Fishing claims to have saved twenty million rubles just from this
service.)

These capabilities are directly applicable to current Soviet technological weaknesses. Their electronics industry, for example, is years
behind Western competitors, but space-based factories could allow the
USSR to overtake Western technology in a very short time. And the
USSR is still comparatively a poorly mapped and poorly surveyed
country; space stations can end this weakness quite cheaply—and decades
earlier than could competing technologies. Additionally, the marginal
agriculture in the Soviet Union would stand to benefit immensely from
improved climate forecasting allowed by space-based observations.

All of these benefits, of course, require that the USSR adopt an
innovative and revolutionary posture toward the opportunities offered
by their space advances. Based on past performance of the Soviet
industrial base, that development is problematical at best—but the
potential for breakthroughs may arouse an uncharacteristic passion for
space-based innovative developments.

Salyut Limitations

However impressive the courage of cosmonauts spending more
than a half year in orbit, the shortcomings of the Salyut system must
not be overlooked. It is still a small station with very limited power
supplies (under 4,000 watts) and little flexibility in replacing equipment.
The station's orbital inclination of 52 degrees does not allow it to pass
over two-thirds of Soviet territory. And the major transportation bottleneck is the return leg to Earth: many tonnes of supplies and crewmen
can be sent up to orbit every year, but the returning crewmen have
only been able to bring back a few hundred kilograms of products and
experimental results from their orbital sojourns.

To escape from these limitations, Soviet space engineers have
embarked on at least two major projects. First, special modules of the
same type as *Kosmos 1443* (linked up to *Salyut 7* in 1983) can be sent
into orbit to provide additional equipment, living space, and electrical
power. Second, a small "space shuttle" vehicle (tested three times in
1982–83) can provide transportation for larger teams of cosmonauts to
and from space, and can presumably carry significantly larger cargoes
than can the present *Soyuz T*. And near the end of the 1980s, a much
larger space booster will allow the launching of space stations for carrying ten or more cosmonauts.

The Soviet *Salyut 6* space station with solar power panels and a *Soyuz* ferry ship docked to one end, viewed from another approaching manned spaceship.

THE AMERICAN ADVANTAGE

The American advantage lies with the space shuttle. Although the NASA spaceplane has not delivered on its promise of cheaper launching costs, it has provided a potentially more rewarding payoff: special services. The presence of people in orbit allows payloads to be built more cheaply and less complexly since human beings will be along to adjust and control them, at least in their earliest (and historically most dangerous) orbits. And the size of the shuttle allows non-astronaut specialists and scientists (including non-Americans) to travel into space to operate their own special apparatus. Such apparatus can be flown and operated, modified on Earth, and flown again within a matter of months. This allows a rapid development of techniques. The weight of scientific apparatus, the electrical power available, and the large amount of products which can be carried back to Earth make the space shuttle

The space shuttle Orbiter, with its payload bay doors opened to expose antennas, radiators, cargo, and instruments, is here viewed from a subsatellite during STS-7. (Photograph courtesy of NASA.)

and Spacelab a far more powerful research facility than the current Soviet Salyut.

But such missions, while productive, are far too short for some of the most important applications. The maximum length of a Spacelab mission is nine or ten days. (With double shifts and specialist-crew-members, that is equivalent to perhaps thirty or forty days of a Salyut mission.) Budget constraints have prevented Spacelab from flying more than twice a year for many years to come, even though many other shuttle flights will be carrying some auxiliary research apparatus on their own missions.

A U.S. SPACE STATION

NASA's current goal is to develop a permanent space station of its own, perhaps to fly by 1991 or 1992. Far from being a copy of the current Soviet Salyut system, it would attempt to be a technological advance and to be at least as capable as anything the Soviets could launch in the same time period. In 1983, NASA specified the capabilities of such platforms—and demonstrated that the American space stations of the 1990s will be several generations ahead of the old Skylab and the current minimal prototype platforms being operated by Soviet cosmonauts.

In power alone, the advance is striking. Skylab used twelve kilowatts of solar power; the current *Salyut 7*, even when augmented by add-on power modules, barely reaches half that figure. The post-1990 U.S. platform is supposed to have at least 64 kilowatts available to the users alone.

In volume, the station will have more than 10,000 cubic feet, about the same as Skylab and twice that of the expanded *Salyut 7/Kosmos 1443* complex first manned in mid-1983. The data rate from the projected station, at least eighty megabits per second, dwarfs any previous manned space facility.

Meanwhile, one intermediate step could be the extension of shuttle/spacelab mission durations to 30–60 days. To do this, solar power panels and maneuvering flywheels must be added to the space shuttle to provide the needed long-term electrical energy and pointing control. NASA officials do not like this approach because it would keep each space shuttle out of service for long periods of time—and there will be customers waiting their turn impatiently. Perhaps another reason is that the availability of such a capability might detract from the arguments for a full permanent manned space station and might delay its development by many years more.

COMMERCIAL EXPLOITATION OF SPACE

Despite the innovative and forward-looking philosophy of American industry, the commercial possibilities of space have not been widely appreciated by private corporations. This is partially due to the long time between initial investment and final profit. It is also due in no small part to uncertainties about trade secrets and reliability of NASA's promised services.

Joint Endeavor Agreements

The one major success in the commercial area has been the Joint Endeavor Agreement (JEA) between NASA and a private team from McDonnell Douglas (the aerospace corporation) and Johnson and Johnson (a large drug company). The private groups built the CFES unit—

Continuous Flow Electrophoresis System—to purify biological materials in space; NASA agreed to fly it several times in space, for free, without prying too much into its actual applications.

The project promises to lead to dramatic breakthroughs in the availability of drugs to treat a number of diseases, such as diabetes, emphysema, dwarfism, thrombosis, and viral infection. During STS-6, in 1983, for example, the 400-kilogram unit was used to process a mixture of rat and egg albumins, a cell culture fluid containing proteins, and two samples of hemoglobin. Researchers found that it was possible to process seven hundred times the quantity of material, with four times the purity, of materials processed on Earth.

On the STS-41D mission (the twelfth shuttle flight) in 1984, McDonnell Douglas had the unprecedented right to send one of its own engineers into orbit to work the equipment. Charles Walker, 35, was thus one of the first non-NASA *payload specialist* astronauts. His presence was justified by a special production run of material to be used in actual clinical trials for FDA approval. McDonnell Douglas wouldn't say what the material is, except that it is a hormone to counteract a human hormone deficiency, and that since it is a "naturally occurring body substance" (i.e., not a new drug) the FDA requirements are nowhere near as time-consuming as for other potential space pharmaceutical products.

That particular product is the subject of the cooperation between McDonnell Douglas, which has built the CFES unit, and Johnson and Johnson, which settled on the exact product. But McDonnell Douglas is making it clear that it will be happy to process drugs for anyone else, restricted only by exclusive per-drug deals with Johnson and Johnson. With the workability of the equipment verified and with bountiful new markets (and vast public health benefits) within sight, several other drug companies have begun discussions with McDonnell Douglas.

Plans are for the CFES unit to lead to a 2,400-kilogram unit (with 24 times the capability of the current unit) to be flown in the space shuttle payload bay beginning in 1985. Commercial quantities of space-processed drugs are to be available the following year. Soon afterwards, McDonnell Douglas would like to attach a modified unit to a permanent space platform.

The apparent success of the CFES project has spurred other corporations to look more closely at space industrialization. The international precious metals company, Johnson and Matthey, has begun investigating eight potential processes for space factories. One is the manufacture of irridium crucibles in space; the higher purity of the units will allow them to be used in the production of higher quality crystals in factories back on Earth. (Johnson and Matthey has not released details of the other processes.) Another material likely to be more perfect if processed in space is gallium arsenide crystals, which cost sixty thousand dollars per kilogram to make on Earth even though an average of 70 percent of most batches are flawed and useless. Eli Lilly, the pharmaceuticals firm, is interested in biological processing to develop entirely new medicines; the Celanese Corporation has a similar interest in new plastics; other groups have begun to examine the creation of biologically active membranes, or the monitoring of hazardous waste. Many such groups are prepared to make substantial investments privately once they are assured that there will be a functioning laboratory, preferably a permanently manned space station, in orbit to provide support services.

Another Joint Endeavor Agreement was signed in mid-1983, with Microgravity Research Associates, a small corporation in Florida. The company will build a small furnace called the Electroepitaxy Crystal Growth (ECG) system, which grows crystals by driving an electrical current through a tube which contains a "seed" at one end while controlling the temperature along the growth region. The furnace will fly aboard the shuttle in 1985. Four additional flights will explore the processes of crystal growth, and two final flights will produce crystals of commercial value.

In 1984 another biological product JEA went into effect, along with a new crystal growth project which probably (but no one is telling) involves a radiation detector material.

Also in the works is a study by Westech Systems Corporation for growing large silicon crystals in space using a "float-zone" process. A silicon rod about four inches in diameter is passed through a furnace where it melts and then crystallizes. Nearly defect-free crystals are the goal, but long flights—several months at least—are required for best results.

With the ball rolling at last, the horizons for commercial processing in space finally look limitless. The aerospace corporation TRW made a study of potential products and identified more than four hundred alloys that cannot be made on Earth, including those "yielding unusual strength or hitherto unrealized mechanical, electrical and magnetic properties." In certain compositional ranges, for example, alloys such as copper-lead or aluminum-lead display self-lubricating properties— essentially zero friction and wear, without maintenance.

The TRW study (carried out for NASA headquarters in April 1983) tried to quantify such possibilities. They produced a tempting array of items:

Semiconductor materials: specific crystal types grown by directional solidification, using 40 kilowatts of electricity, producing 370 pounds of material in a ninety-day run. Commercial value, $140 million per year;

Vapor-grown crystals: small quantities of material produced by vapor transport, using 30 kilowatts for a thirty-day run producing 165 pounds per run. Commercial value, up to $600 million per year;

Solution-grown crystals: specific crystal types grown by solution transport, using 15 kilowatts for a ninety-day run producing 700 pounds of material. Commercial value, up to $650 million per year;

Commercial manufacturing of glass, using containerless processing: ninety-day process using 12 kilowatts, producing 2,600 pounds per run. Commercial value, one billion dollars per year.

Biological and pharmaceutical processing was also mentioned in the TRW study but without specifics. However, another NASA survey provided the following astounding list of potential lifesaving space products:

Anti-hemophiliac Factor (AHF): Over 23,000 hemophiliacs in the United States are treated twice weekly. Impurities in drugs cause allergic reactions and often block effective treatment. Space-processed AHF could be thirty times purer and twice as efficient, thus cutting the needed raw blood supply in half.

Erythropoitin, a drug crucial to those suffering from chronic anemia or kidney dysfunction, is normally produced by healthy kidneys to stimulate red blood cell production. It cannot presently be produced commercially, but it is estimated that one million Americans could benefit from the space-based refining of only one pound of erythropoitin per year.

Beta cells make insulin in a healthy pancreas. If processed in a pure enough form, to resist rejection, they could be injected into a diabetic's liver, take root, and cure the patient's diabetes. There are three million people who could use this space-processed material.

Urokinase is used to dissolve blood clots. By preventing deaths from thromboembolisms, a commercial supply could save fifty thousand lives a year.

Antitrypsin products, valuable in cancer chemotherapy, could help an estimated half million patients per year.

The enzyme alpha trypsin is used to treat emphysema.

Collagen is being studied by the Battelle Labs in Columbus, Ohio. It is a fibrous insoluble protein found in connective tissue (which comprises about 30 percent of the body's weight). If pure enough, it can be used to produce artificial corneas, replace damaged blood vessels, and treat severe burns. Space experiments are planned to separate the effects of gravity from other factors that control cell growth, leading to the discovery of ways to control the shape and character of collagen fibers so that biomedical materials—human implants, in particular—can be grown tailor-made.

Blood components such as T- and B-lymphocytes aid in the prevention of organ transplant rejection.

There are products called macrophages which in pure form are useful in the early detection of immunological reactions.

Interferon has frequently been mentioned in the Soviet press as a potential product from Russian space factories.

The Space Systems Division of General Electric is studying the manufacture of tiny latex microspheres in space, for medical applications. They are used in blood-flow experiments and to calibrate medical instruments. Spheres in the 10-micron size range can be sold for thirty thousand dollars per pound but cannot be mass-produced on Earth. Accuport Laboratories, of Huntsville, Alabama, is also studying the commercial production of such microspheres.

These semi-magical products are, of course, only one aspect of producing wealth and health from space-based industries. Estimates have been made that information collected in space—particularly surface photography—could be worth a quarter of a billion dollars annually by 1990 (so says a 1982 report by Terra-Mar Associates in California). For example, the St. Regis Corporation is actively studying methods whereby the health and size of trees in the company's forest lands can be monitored and inventoried from space.

And even a topic as far-out as "space tourism" may not be totally out of reach: a TRW study indicated that seats sold at $200,000 each would find a ready market, and by the mid-1990s several dozen tourists could be riding on "space available" seats on advanced space shuttles, every year.

All these activities would most productively be occurring aboard an American space station, or on automated "space platforms" tended occasionally by visiting astronaut crews.

CONVERGING PATHS

These observations suggest that the space applications paths of the United States and the USSR may soon start converging again. The United States will seek to develop a space station while efficiently utilizing the advantages of the space shuttle. Conversely, the USSR will seek to improve its space transportation system with winged spaceplanes and a bigger booster while taking maximum advantage of the small permanent space station it is now operating.

Each side, in other words, will be seeking to develop capabilities already mastered by the other. This startling situation offers generally unrecognized opportunities for both third-party contributions as well as significant joint endeavors by the United States and the USSR. Such potential may, in fact, be the most valuable product of current space developments: at a time when diplomacy is pulling the world apart, space technology—and a common desire to maximize benefits to be derived from space activities—can offer a surprisingly powerful argument for closer international cooperation.

Guest Cosmonauts

The most visible manifestation of this trend is the presence of "guest cosmonauts" aboard Soviet space stations and of European payload specialists aboard space shuttle missions. Their presence could, at first, have been merely for show, but the near future's reality may soon surpass the original image making. As with the recent flights of women into space, it is another case of the right action for the wrong reasons.

For the Soviets, the participation of one representative of each main Soviet-bloc nation was an easy and cheap way to make political propaganda. Since the missions had to be flown to support long-duration crews in any event, little additional cost was incurred by placing non-Soviet "cosmonaut researchers" aboard. The political aspects of the program were made clear by the fact that the second batch of six nations was launched in strict Russian Cyrillic alphabetical order, obviously without the slightest regard for technical or scientific factors. This program (which seems to have been instigated by the announcement in 1977 of the original Spacelab agreements calling for the flights of European astronauts "as early as 1981") continued in 1982 with the flights of a French "spationaut" and of a woman cosmonaut; both were assigned the "guest" seat aboard the *Soyuz T,* with minimum responsibility for the actual execution of the mission. An Indian guest flew in 1984.

Although conceived as a stunt, the guest cosmonaut program opened the door for genuine space research by participating nations. Several of them took advantage—particularly East Germany, France, and Hungary—and developed instruments for general use aboard Salyut, even by other nation's representatives. The French research apparatus was particularly impressive.

Aboard the space shuttle, European researchers have found many opportunities for authentic contributions of their own. The *SPAS-01* spacecraft, built by a German aerospace firm, Messerschmidt Bölkow-Blohm, and flown on the STS-7 mission, is one good example. Spacelab itself is another—both the first mission and the dedicated German research flight, *Spacelab D-1*, now set for late 1985 with several European crewmembers.

So European space researchers (and Japanese as well, along with anyone else interested) can attach their space research apparatus to large manned space vehicles of the United States and, quite possibly, the USSR as well.

A Space Handshake

The next step is for the United States and the USSR to swap some apparatus for space research, each nation taking advantage of unique capabilities of the other. Currently, the Soviets can keep manned platforms in orbit indefinitely, while the United States cannot; the Americans can carry large loads into space, *and back,* which the Soviets cannot. Technical requirements make it logical for each of these two mutually complementary programs to take advantage of what the other side has to offer. And while the current diplomatic situation makes such a new "space handshake" seem like naive daydreaming, there are in fact political benefits, in terms of international and domestic politics, to be gained for both nations. A detailed proposal for such a joint mission can be found in chapter 12.

The United States and the USSR would not need to merge their programs, of course. That is both impractical and unnecessary. But a fraction of the space research budgets of both nations, along with the technological capabilities of third parties, could allow the accomplishment of some spectacular feats which neither nation would ever probably have considered on its own.

And among these large projects will be the contributions of other nations. Spacelab modules might, in the early 1990s, be attached to both Soviet and American space platforms.

A decade from now, even more far-reaching projects can become feasible. Already, studies have shown that an "oxygen mine" on the Moon could be commercially profitable by the turn of the century, by importing liquid oxygen (baked from lunar soil) to orbits near Earth, where it would be used to fuel the transportation of very large communications and power satellites into 24-hour *geosynchronous* orbits (in which the satellite remains still in relation to a point on Earth's surface). The construction of such very large structures (such as antennas for Earth communications, or for microwave observations of the dynamics of the atmosphere, or for the search for extraterrestrial intelligence) could make the mining of lunar metal a profitable industry. From there, it would not be long until near-Earth asteroids become a source of metals both for space construction and—ultimately—for import down to Earth itself.

THE NEW RACE

It is to reach such goals, and the numerous closer but equally profitable ones, that both the United States and the USSR are racing so hard in space. That race, while not as easily measured as the head-to-head competitions of the early Space Age, is for even bigger stakes. Because of this race, unique and highly promising opportunities have been opened for the full-scale participation of any nation in the world willing to make a down payment in anticipation of the future payoffs from such activity. Those nations and groups of nations which are doing so will enrich both themselves and the rest of the world by their farsighted investments.

In a larger sense, the race is, as H. G. Wells said long ago, "between education and catastrophe"—in modern terms, between human technological abilities and wisdom, and potential disasters both man-made and natural. The wealth and wisdom conceivably available from space will not merely be applied to frivolous luxuries, but may be crucial to the survival of human civilization on (and off) this planet. In that sense, we are all in the race together. And we should indeed be running very, very hard.

[handwritten marginal notes: "Military Security", "econ stakes", "are liberalist idea? ⟹ but, still realist². Security concern?!"]

2

Spaceflight Fire

When Apollo astronaut Michael Collins wrote his marvelous book about his spaceflight career, he called it *Carrying the Fire*. The metaphor was supposed to signify the launching of manned rockets into space. Such an action is conducted very carefully, just as if one were carrying fire. And the return from space is a baptism of fire hitherto experienced only by meteorites.

PERSONAL FIRE

I have seen the fires at both ends of the mission—launching and entry—and I cannot say which of the two is more thrilling to watch in person. In neither case was I at all prepared by earlier televised versions, or by my technical familiarity with what was actually going on.

The first time I saw a manned space launch was in 1968 when, as a college student, I spent my Christmas vacation hitchhiking to Florida and back, in order to be present for the blast-off of mankind's first voyage to the Moon. I was there, camped out on a causeway, when

Apollo 8 and its three-man crew leaped into the sky atop a Saturn 5 booster. My first impression was of the pad silently engulfed in a giant white cloud of steam, which for long, agonizing moments seemed to be all we might ever see. Then the white obelisk rose slowly from the top of the cloud, the reflections of the flames making a scintillating roadway running across the water from where I stood in relation to the pad. Only later, with the reverse-meteor climbing high in the sky, did the rumble and crack-crack-crack of the engine noises reach us. Through binoculars I watched the first stage fall away, still high in the sky to the east, the man made flames of the *Apollo* soaring to merge with the blinding brilliance of the morning sun. It was damp and cold that midwinter morning, but that did not explain all the chills running along my spine.

The second time I was an eyewitness to spaceflight was in March 1982 after I had been transferred to a new control team which was not supporting the third shuttle mission. I was the guest of a national television network, there to put the space shuttle into perspective with the Soviet manned space station program. During the long network coverage of the countdown, I provided some insights into the differences and similarities of the goals and technologies of the two competing spacefaring nations. Since I was no longer needed as the count reached zero, I took my place by the press bleachers and prepared to watch.

Again, there was a burst of steam clouds—but with a difference. The solid rocket motors cast a lurid flame from inside, and as the still-hidden spaceship approached the top of the shrouding clouds, the colors took on a new incandescent glow for an instant. Then the spaceship, clearing the cloud, pierced upwards with an impression of power far greater than my memories of a decade and a half earlier.

And the flames! It seemed like a sliver of the midmorning sun, already high in the bright blue sky, had been broken off and carried by some magic to be attached to the tail of the spaceship. It was dazzling in brightness, too bright even to look at—without tears welling up.

Although the noise was more familiar, it was much enhanced (I was much closer this time). Nothing at all related had ever come over the sound systems of the televisions and radios on which I had monitored earlier blast offs. The full range of sounds, and of feelings against the face, chest, and through the soles of my feet, was startling. You do not watch a space shuttle launch—you experience it.

For the first space shuttle landing in Florida, in February 1984, Americans got their first chance to witness the flames of atmospheric entry. The spaceship's flight path took it due south of Houston, about seventeen minutes before touchdown a thousand miles farther to the east. The flight plan showed it would be forty miles high, doing nine thousand miles per hour. It would still be in the "blackout" phase, surrounded by a radio-deadening superheated ion sheath, but by the time it passed south of New Orleans a few moments later it would have resumed radio contact. Geometric computations showed it would pass about thirty degrees high in the southern skies, from right to left.

I did not know what to expect that predawn morning, with only a faint glow of the coming sunrise, to the east, to my left. Upon going outside, my first impression was of frustrated disappointment—there was almost total overcast. But on careful observation, a few stars appeared, and I could make out Scorpio to the southeast and Venus and Jupiter to the left of the constellation, popping in and out behind passing wisps of clouds.

The clock told me that the shuttle would be approaching, but I had no idea what it would look like. A small meteor with a flaming tail? A bright light, reddish in tinge, lit by the Sun just over the horizon? Nothing at all?

It turned out to be none of those things.

The clouds to the southwest were thinning, but still no stars were visible. Suddenly a glow appeared there, like dim moonlight seen through thick overcast. Something—the spaceship!—was blazing its way across the skies, but hidden from my view.

Then it broke into a small clear area directly south of me. No meteor, or starlike point of light—the apparition was a brilliant milky-white line appearing magically across the black backdrop. No object appeared at the apex—the line just sprang into existence and did not fade. Like a supernatural finger leaving a trail of phosphorescence, the line crossed the southern sky and then lingered, fading without widening, for half a minute or more.

I remember jumping up and down and shouting, "Come on home!," and being wrapped up entirely in the visual wonder. But it didn't take me long to become an engineer again and ask myself, "Just what the hell caused THAT?" I listened intently for any sonic boom, but heard nothing.

It was unlike any fireball I had ever seen, and I've seen a bunch of weird ones. The flames lingered too long, and the spaceship itself was not engulfed in its own flaming sheath. Later, when I examined photographs taken of the apparition by my friend Paul Maley, it became clear that the trail was some sort of chemo-luminescent residue emitted by the spaceship—probably exhaust plumes of attitude control rockets in the tail, firing sideways to keep the ship's nose pointed correctly.

Ammonia fumes given off by the hydraulic power units (which drove the wing surfaces) may also have contributed.

As far as we could discover, Paul and I were the only space engineers in Houston who witnessed the flame trail. Around Clear Lake, where NASA is located and where most engineers live, the overcast was total. I live twenty miles to the south, more than halfway to Galveston, and Paul drove himself to the Galveston Island causeway to set up his cameras. Our acts of faith and hope were rewarded by a small—but perfectly placed—hole in the cloud cover, through which shone the flaming glory of the spaceship's streaking return to Earth.

In the years ahead, other shuttle missions will on occasion land at night or in very early morning in Florida, and other people will watch for—and be rewarded by seeing—the spectacle. But I have a personal stake in the project, and that morning I got a very exclusively personal reward.

SPACE FIRE MACHINERY

The machinery designed to "carry fire" must be the most powerful and reliable parts of a manned spaceship. Disaster can—and occasionally does—occur with lightning speed.

First there was the embarrassing problem of the loose thermal tiles for reentry protection. With that solved, and the first few flights successfully accomplished, a new embarrassment arose: the main engines of the second space-qualified shuttle, the *Challenger,* leaked repeatedly during launch preparations. Blast-off slipped for months.

The two troublesome technologies share one feature: fire. One protects against the flames of entry into Earth's atmosphere, as tremendous orbital potential and kinetic energy is bled off as heat; the other generates heat to create kinetic energy for expelling reaction mass and propelling the spaceship high and fast. More than merely metaphorically, the flames balance each other out. And the technological embarrassments were similar as well.

THE SPACE SHUTTLE MAIN ENGINES

It is a surprise to many viewers of space launchings that the flames from a hydrogen-fuelled engine are practically invisible. The Solid Rocket Boosters (SRBs) leave searing streaks in the sky—"a piece off the face of the sun" is how I instinctively thought of it when I witnessed my first space shuttle blast-off—but the powerful exhaust of the Main

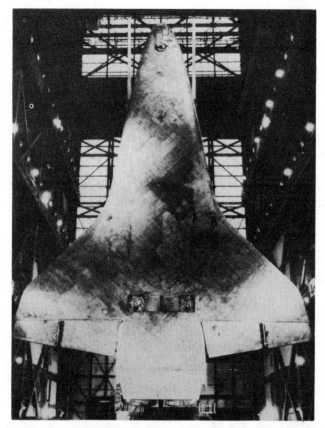

On entry into the atmosphere, the Orbiter rides a plasma wave as hot as the Sun's surface, but tiles insulate the spaceship's structure. The pair of holes near the bottom are for propellants which flow from the External Tank into the Orbiter's three main engines; after the tank is jettisoned, the doors snap closed to insure safe atmospheric entry. (Photograph courtesy of NASA.)

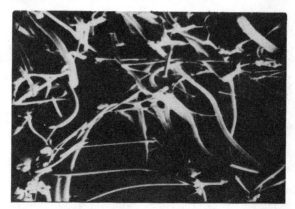

Enlarged view of shuttle tile shows mostly empty space between filaments—the best insulator is a vacuum! (Photograph courtesy of NASA.)

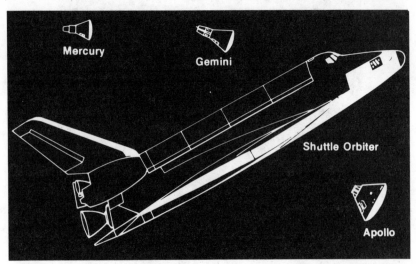

Comparison of size of space shuttle to earlier space capsules hints at immense problems that needed to be mastered in flight control and thermal protection. (Photograph courtesy of NASA.)

Engines is barely visible. The flames of reentry are supposed to be well known, but when Dale Gardner looked over his shoulder out the over-head window on STS-8 and saw flames *above* the shuttle, too, a shock went through him—and through the engineers who later watched his eerie film of the tongues of flame over the spaceship's spine.

Few observers realize the enormous technological advance in space propulsion which is represented by these engines, known in NASA parlance as SSMEs (Space Shuttle Main Engines). In many ways the entire space shuttle does not appear particularly difficult considering the miraculous missions of the Apollo and Skylab programs.

A news media eager to highlight problems has not helped the image much, either. For example, despite what you may think, there has never been an explosion of a shuttle engine during any testing. There have been failures and bad fires (one engine was completely destroyed that way) but the spectacular blowups that marked the development of the Saturn 5 engines have no equivalent in the SSME program.

The SSME Versus the J-2

A comparison of the SSME and the Saturn 5 upper-stage engine, the J-2, can be instructive. Both use liquid hydrogen as a fuel and liquid oxygen as oxidizer—so-called cryogenic propellants which the Soviet

Union has never managed to master. Both engines took about seven years to develop from start to first manned launch.

The efficiency of the SSME is about 10 percent greater than the J-2. Its thrust is about twice as great but it weighs considerably less, giving a power-to-weight ratio three times better than the J-2 (and two hundred times better than an automobile engine). Part of this improvement is in the fact that the combustion chamber pressure of the SSME is four times as great as that of the J-2.

There is no comparison in lifetime required. Each J-2 was used once, for several minutes during a single stage of a three-stage launch. The design goal for an SSME is 7½ hours, enough time for fifty-five flights, and it burns from the moment of lift-off until the shuttle reaches orbit.

The J-2 was turned on to full power and stayed that way. The SSME must be throttled up and down. This can be as low as 65 percent during delicate portions of the aerodynamic ascent and to avoid overstressing the vehicle when most of the heavy propellant is exhausted. During emergency maneuvers, the engine can be pushed up to 109 percent. Within a few years, each SSME will be operating routinely at 109 percent of "rated thrust" (512,000 pounds), with emergency levels of 115 percent. The engine may someday be redesigned to provide 130 percent of the current power levels.

A special new alloy was developed for the thrust chamber (where temperatures exceed the boiling point of iron). Called Narloy-Z, the metal is basically copper with small amounts of silver and zirconium. It is then plated with several layers of nickel. Weight limitations forced a heavy emphasis on welding of parts, using electron-beam welding techniques. There were difficulties encountered here, all the way through the manufacture process and (to NASA's embarrassment) even on the launchpad itself!

About thirty of these engines will be built by the Rocketdyne Corporation. The first three flight models, from the *Columbia,* are now being rebuilt for installation on the *Atlantis* in 1985. Three new engines were provided for *Challenger,* but one was replaced (twice!) with other new engines originally scheduled for *Columbia* when it took to space again in late 1983 with Spacelab 1. Yet another three new engines were built for *Discovery,* which made its maiden orbital excursion in 1984.

During the J-2 program, three engines failed in flight; the only manned launch this happened on was *Apollo 13.* It is hoped that the SSME program will have an even better record and that no failures will ever occur. However, if an engine does fail during launch, there are emergency procedures. They are desperate ones, to be sure, but the alternatives are worse.

THE FIRST SHUTTLE LAUNCH

I have a tape recording of the voice channels in Mission Control when we first launched a space shuttle. When I listen to it, I'm glad I was there so that I can identify the voices of my colleagues. Otherwise, it would be hard to determine their identity since all the voices seemed pitched so high—as high as the tension which filled the room. On April 12, 1981, when *Columbia* first took to the sky atop three SSMEs and two flaming SRBs, we really did not know what would happen.

Ready for Armageddon

My role in the mission was to man the console called Ascent Consumables. It supported the Prop console (maneuvering and altitude control rockets) in the "front room," the section of the Mission Control Center seen on television. The job involved monitoring the countdown and launch, and then handing over to another engineer for orbital operations—before returning for the night shift when the crew was asleep. Yet another team of specialists operated during the return to Earth and landing phase.

As it turned out, of course, the ascent—the uphill climb as we called it—went perfectly. Almost perfectly, that is—when the astronauts forgot to activate one switch, the Flight Director had become relaxed enough to joke about it. "Charlie," he said to the flight controller responsible for that particular system, "the first crew that remembers to turn this switch is going to get a medal." There was no loud laughter, but there was widespread relaxation.

Afterwards we all got letters of appreciation. The Flight Director, Neil Hutchinson, congratulated us on the alert, eager way we had trained for the mission over the previous two years. Things had indeed gone smoothly, but we were, in the words of our Flight Director, "ready for Armageddon"—ready for any potential disaster.

Simulated Disasters

We had to be ready for anything. So many things might have gone wrong. So many different plans had been drawn up, analyzed, practiced, refined, and practiced again. If one failure occurred, do this. If another failure occurred, do that. If both should occur, do yet a third procedure. And in the event of an unanticipated failure (which after all was most to be expected), be prepared to think fast and accurately, within a space of seconds.

Many such multiple failures were thrown at us a few months before launch. One was, in fact, thrown directly at me. With several fuel tanks leaking simultaneously, I suddenly realized that the crew would have to do exactly the opposite of the planned procedure for a single leak. I passed the word to the Flight Director, "Do not, repeat, do not close the isolation valves—let the tank pressure bleed down to create ullage volume." The logic was persuasive, and the crew was quickly told. Crippen, the copilot, took a moment to understand. "Are you absolutely certain?" he demanded, as he reached for the switches. My heart sank since there wasn't time for an elaborate explanation. Then he understood, "Of course, I see why!" And the men managed to return safely to Earth.

Sometimes the simulated failures were not survivable. One occurrence was every spaceman's nightmare come true: all the stored energy intended to carry their capsule on an eight-minute trek into orbit released catastrophically in a few seconds. Both the crew and Mission Control would be helpless in such a situation, but there was one "ace in the hole."

When only two men flew the *Columbia* and the *Enterprise*, they had the option of ejecting from the spaceships near launch or landing if the vehicle went out of control. This illustration shows a test firing at White Sands. The seats have since been disarmed and will be removed; other shuttles never had them. (Photograph courtesy of NASA.)

For just such an eventuality on earlier space missions, rocket engineers had built special emergency equipment. The Soviet Vostok capsules (1961–63), American Gemini capsules (1965–66), and all two-man space shuttle flights (*Enterprise* in 1977 and *Columbia* in 1981–82) had ejection seats, designed to catapult the crewmen out of the endangered spaceship where they could descend safely by individual parachute. For American Mercury capsules (1961–63) and Apollo spacecraft (1968–75), and for Soviet Soyuz (1967–present) and lunar Zond (1967–70, never manned) spacecraft, the escape tower system was employed to pull the entire command module free of its booster.

REAL RISKS

In 1964–65, the Russians launched two Voskhod capsules without any launch escape system. Political pressures for a "space spectacular" had forced rocket engineers to take risky shortcuts. Rationalized one of the cosmonauts, "It was a very reliable booster." But many years later it was exactly the same kind of booster which exploded, in September 1983, with two men aboard (see Chapter 10).

Then, late in 1982, the Americans took the same risk: they launched astronauts without an escape system. The space shuttle *Columbia* had finished its flight tests and was to become operational. Instead of a crew of just two test pilots, the spaceship would now carry four or more crewmembers. That is too many for ejection seats, and the shuttle vehicle is too large for an escape tower. So in case of launch disaster there is no way out.

There are some emergency procedures, of course. Now the entire spaceship and its large crew must be recovered intact.

This seems excessively hazardous at first glance. But it is, after all, how commercial aircraft are operated. And since the space shuttle is supposed to bring routine service to space, extraordinary safety measures may no longer be justified.

Plans now call for the shuttle to make an emergency landing if trouble develops in the engines during the eight minutes of launch. Early in this sequence, the spaceship can turn around and return to Florida. Later, the spaceship can push onward to reach a landing site across the ocean (usually Dakar in Senegal). Still later, it can circle the Earth once and reach an emergency airfield in the western United States. If an engine failure occurs late in the ascent, the spaceship can limp into a low, barely stable orbit, for several hours.

There are, of course, conceivable failures which would not allow such emergency recoveries. One of the solid boosters could explode,

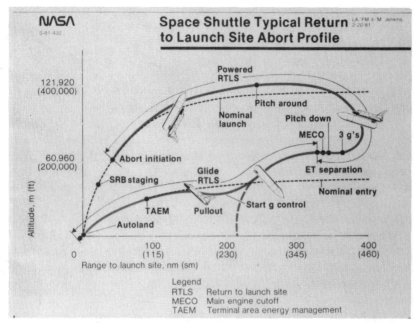

For various launching failures, the shuttle is to turn around and perform a Return to Launch Site, or RTLS. (Photograph courtesy of NASA.)

or the external fuel tank could disintegrate, or all three main engines could fail. In that case, as one astronaut once blithely remarked, it would be curtains—the end of the act.

That remark was made by Daniel Brandenstein, the copilot on the STS-8 flight in August 1983. NASA later announced that the nozzle of one of the two solid rocket boosters had come within a few seconds of burning through, an eventuality which could have brought disaster. The dramatic announcement received immense news media attention.

But it was an overreaction. After all, the nozzle had worked. It had been designed to withstand the tremendous temperatures for two minutes under the worst possible of conditions—and it had done so. The conditions had been worse than expected, but that is exactly why the spaceship's equipment is designed from the start to be strong enough to tolerate far worse conditions than are expected.

Had the nozzle burned through, it would have been very near the end of the planned burning sequence—the solid boosters were about to be dropped off anyhow. Perhaps a small sideways force would have resulted, but the flames would not have reached the main fuel tank since it is located a great distance from the nozzles. The other engines could have been tilted on computer command to make up for the deviation which would have lasted only a few seconds at most.

Thus there never was a "near catastrophe," as some hysterical reports claimed. To be prudent, the nozzles on the next mission were replaced for more thorough manufacturing—but they would almost certainly have functioned adequately without any action.

The most amusing aspect of the excitement over the STS-8 nozzle problem was the way it was reported in *Pravda*. There, in Moscow, the story received more attention than the original flight had. The headline read, "An Explosion Could Have Occurred." But of course it was in Russia that a manned spaceship had really exploded only a few weeks after the STS-8 mission—an actual explosion which received not a single line in the official Soviet news media!

Still, no feelings of smug superiority could be tolerated in Houston. Launching manned spaceships is a dangerous business—as recent Soviet and American experience amply illustrates. The only choice is to hope for the best but to prepare for the worst.

FLAMES OF REENTRY

During atmospheric entry, a spaceship may seem, for all practical purposes, to be skimming the surface of the Sun for several minutes. The heat is not—as is commonly misbelieved—due to "atmospheric friction." Instead, the physical process involves *compression* of air molecules ahead of the vehicle as it snowplows through the thin atmosphere. In their panic to get out of the way, the gas molecules become superheated and compressed and turn to plasma as electrons tear loose from atomic nuclei. The result is a constantly renewed mass of 10,000°F superhot gas, several feet in front of the spacecraft. Heat reaches the vehicle via radiation from this zone. To awestruck shuttle crewmen peering downward out the front windows (a view never afforded to riders in earlier "space capsules"), the salmon-red glow looks like "the inside of a blast furnace"—or the mouth of hell.

The ionized gas, being an electric conductor, is opaque to radio waves. This causes the well-known "blackout" phenomenon. Since this plasma sheath may not extend all around the spacecraft, it is hoped that radio communication can be re-routed via a relay satellite in space high above the shuttle. But this requires precise preprogrammed aiming and tracking by the relay satellite's swivelable dish antennas, and initial attempts in 1983 were not successful.

LOW-EARTH ORBIT—HALFWAY THERE

Space enthusiasts have often expressed disappointment about the fact that the space shuttle will only fly in near-Earth orbit—a far cry from the Apollo expeditions and the dreams of human voyages to other planets. Low-Earth orbit (LEO in space acronym) is viewed by some as unworthy of the term spaceflight. It seems so limited, barely scratching the surface of the universe and hardly worth the risk of fire.

But pessimism about our growing space capabilities is uncalled for. Don't sell the shuttle short. In a phrase attributed to Robert Heinlein, "Once you're in orbit, you're halfway to anywhere." To that I'll add my own corollary: "And once you're in orbit, you're no longer in such a hurry. You can take your time and turn on some engines that are *really* efficient."

It's all in the numbers. The velocity of a spacecraft orbiting 100 to 200 miles above our home planet (about 25,000 feet per second) is already 70 percent of the velocity needed for complete escape from Earth's gravity. And that's the hardest fraction since it has to be accumulated in a matter of minutes to counteract the incessant tug of gravity on the climbing spacecraft.

Once the spacecraft is traveling fast enough to be in a safe, stable orbit (Earth's horizon falls away beneath it at the same rate as its path curves downward), it can "shift gears" and use radically different propulsion systems that may need hours or days to build up speed. These might be systems which could never lift off the planet's surface, but in space they can operate with a fuel efficiency many times greater than that of the high-thrust chemical-burning engines needed for the initial frantic uphill climb.

To appreciate the significance of these different types of rocket propulsion systems, we need a few simple engineering terms. Rocket efficiency is measured as *specific impulse* (Isp) in seconds. There are some esoteric engineering explanations for Isp, but think of it this way: An engine's specific impulse is either the time period in which one pound of propellant provides one pound of thrust, or the amount of thrust which one pound of fuel can maintain for one full second.

For example, an engine with an Isp of 300 seconds and a rated thrust of 30,000 pounds would consume 100 pounds of propellant per second; over a period of five minutes (or 300 seconds), the engine would burn 30,000 pounds of propellant.

The thrust is mainly a factor of the engine's design (engine thrust capability varies from only a fraction of a pound to more than a million); the specific impulse depends on the chemistry of the propellants, the pressure and temperature of the fire in the nozzle (usually, but not necessarily, the product of combustion), and the ejection velocity of the propellants. But that's a subject for a graduate-level college course in mechanical engineering, and for a later chapter (see Chapter 17).

Here are some typical efficiencies: The kerosene-liquid-oxygen engines of standard space boosters have Isp values from 250 to 300 seconds. Liquid-hydrogen fuel provides an Isp somewhere in the 400 to 450 range. Nuclear engines (such as the now defunct NERVA project, which died for lack of a target) have low thrusts but could achieve 800 to 1,500 seconds of Isp. Ion engines have very low thrust but can operate for months with Isp values above 10,000 seconds. The semi-fictional matter-antimatter drive would have values above 100,000 seconds. Solar sails wafting on sunbeams could effectively have a specific impulse approaching infinity (although they take an awfully long time to build up any speed). Mass drivers (electromagnetic catapults) promise very high efficiency—if they can ever be assembled in space.

My point in reviewing this hardware is that getting into orbit uses up at least 70 percent of the fuel needed to leave the planet behind entirely. With more efficient in-space engines, that proportion could (in practice) go as high as 95 to 99 percent. Enter: the shuttle, and the opportunity to develop and utilize the high-efficiency space engines needed for that small additional push required to roam the Solar System—*once* we are in orbit.

Consider the possibilities: First, remember that the geography of the Solar System will not be measured in miles or astronomical units but in feet per second (fps) and in days—that is, velocities needed to proceed from place to place and voyage duration. The *tens of millions* of miles between planets should have no more significance to us than did the *thousands* of miles to Columbus, Magellan, and Marco Polo or the *hundreds* to Odysseus, Jason, and Moses.

To the Moon

If you consider the numbers in any astronomy text, it may *seem* obvious that the Moon is much closer than is Mars or Venus. But that's only if you are still thinking in obsolete units of measurements. For the true scale of the Solar System, reprogram your thinking to realistic travel terms: velocity and time.

Let's start in LEO—100 miles above Earth, where the shuttle can take us—and add up the velocities required to get to the lunar surface and back: escape from LEO, 13,000 fps; injection into lunar orbit and braking to a soft touchdown, 7,000 fps; launch and injection back toward Earth, 7,000 fps. Additional fuel would be needed to brake into LEO again, but it's more economical to use Earth's atmosphere as a cushion to slow the craft down to manageable speeds. That's a total of 27,000 fps beyond the 25,000 fps we already have in LEO.

To Mars

Now let's figure a voyage to Mars. We need an added impulse above and beyond simple escape velocity from LEO (13,000 fps) in order to set us on the outward trajectory from Earth orbit to that of Mars (for the most economical path, the timing must be just right so that Mars is at the same point in its orbit when we arrive—that is possible for about a month every 2½ years). That figure is about 10,000 fps. Once we arrive, how much velocity is needed to make a soft landing? Surprisingly little because, unlike our Moon, Mars has a significant atmosphere that can be used for 98 percent of the braking. Takeoff and return require another 10,000 fps. Total beyond LEO: maybe 35,000 fps at most.

Some Perspective on Distance

The bottom line is that, in terms of velocity, Mars is only slightly more distant than the Moon! In terms of time, of course, it is much farther—many months versus only a few days. So the engineering challenges lie in long-duration life support rather than in propulsion. And significantly, the Soviets are concentrating most of their new manned space research and development in this area.

Conceptually, even that time-distance factor can be conquered with the right kind of hardware. A useful and startling arithmetic exercise is to conjure up a super space drive with high thrust and fantastic specific impulse (say, several hundred thousand seconds) that would allow the vehicle to constantly accelerate at 32 fps (equivalent to one-g, the force of gravity on Earth's surface) for hours or days on a single tank of fuel. How would the geography of the solar system look to an astronavigator on such a spaceship?

The Moon would be less than four hours away, Mars as little as two days, and Pluto a few weeks. Of course, incredible midcourse velocities would result, and the last half of the flight would be spent decelerating.

I don't have the foggiest notion *how* such engines would be built (although I'm blithely confident that the universe is bristling with energy sources which haven't even been noticed yet), but I see no reason why these engines could not be constructed, whether in a century or in a millennium. And engines which can provide a few percent of that power are conceivable today (see Chapter 17).

But for now, just as ancient explorers, traders, and colonists spent years in fragile boats or on dusty caravans, we'll have to creep laboriously from planet to planet. The distance that Odysseus traversed for ten years now takes an air traveler less than a day; the distance that took Magellan's ship three years can be covered in a few hours by a space traveler. So the inner Solar System today is on the same scale to us that the eastern Mediterranean was to Odysseus 3,000 years ago, or the whole world to Magellan 400 years ago. Then, as technology improved, the real distances (and efforts and risks required) diminished by orders of magnitude.

We are again at the beginning of that distance-compression cycle. Our permanent bridgehead into orbit, the space shuttle, will provide the opportunity for us to set sail and continue our outward-bound voyage. And the hardest physical part of that journey is already behind us.

3

Astronauts Versus Robots

The old conflict about the respective values of manned versus unmanned space exploration continues, and it shows that nothing has
been resolved regarding this question. The same arguments simply
recycle themselves every few years, and any space shuttle problems
stir up bitter new confrontations.

Another installment appeared a few years ago in a political commentary magazine widely distributed in the eastern United States. Although it was a city magazine, because of its distribution in the District
of Columbia and environs it had political influence all out of proportion
to its circulation. In a lengthy article in the April 1980 issue, author
Gregg Easterbrook (my childhood dentist's son—I *believe* in the concept of a "small planet"!) wrote that "the Shuttle simply can't do
anything the old rockets couldn't do, won't save money, and won't
help us learn anything we couldn't learn with probes on the old
rockets. . . . Anything we need to know about things near Earth where
the Shuttle can go, we can launch an old-fashioned rocket to discover."

Astronauts are useful only for show, Easterbrook argues. "In space
the men sit and watch the dials. Human judgment is needed only when

the Shuttle has to be flown home"—and that, according to Easterbrook, is a complex process only because of the massive extra weight
of equipment made necessary by the presence of the men in the first
place! The article quotes approvingly a statement by Harvard physicist
Albert Cameron: "It sounds unpleasant, but it's true . . . in space man
is in the way. There is very little work that requires a pair of hands."

This antiastronaut theme was gleefully picked up by a well-known
cynic-at-large, Nicholas Von Hoffman, in a nationally distributed newspaper column. Basing his attitudes on Easterbrook's article and on his
longtime antispaceflight biases, Von Hoffman concluded: "There is
nothing a manned space shuttle can do that a computer-commanded
robot can't do better and cheaper. Wasn't the whole idea of swinging
down from the trees and becoming toolmakers so that machines could
do the dull, difficult, and dangerous work for us?"

In the past, objections to the high cost of manned spaceflight have
come from certain specific groups: Scientists, whose laboratories are
(in their own estimation) underfunded, see the expenses of manned
spaceflight as draining off money that would otherwise have gone to
them; political activists of the left see manned spaceflights as draining
off money which would have been distributed by them—and they have
never forgiven Apollo and Skylab manned space shots for the unpar-
donable sin of making Americans proud of their country; other ideo-
logical spokesmen see costly manned space shots as diverting high
technology from desperately needed space-weapons systems.

Perhaps it is possible to consider the arguments on this issue dis-
passionately. Perhaps—but it has rarely, if ever, been done. Let's try.

PROBLEMS WITH REMOTE CONTROL

The very first point to keep in mind is that there is simply no such
thing as unmanned spaceflight. There are always people, men *and*
women, involved in the control of space vehicles and the analysis of
data received from them. Sometimes a few of these people are actually
aboard the space vehicles and are supported by large teams of specialists back on Earth. At other times the probe is entirely remote-
controlled. In either case, sophisticated computer systems on board
the space vehicles are usually available to handle routine data processing and spacecraft-control functions.

There are definite problems in the remote control of a spacecraft.
First is the problem of *bandwidth,* which restricts the amount of data
about the spacecraft's environment that must be returned rapidly to
the controllers back on Earth. In deep space the volume of such signals

Spacewalking astronauts can service unmanned satellites and enhance their performance. (Photograph courtesy of NASA.)

may not be particularly large because the environment is relatively simple; near a complex planet or on the surface of an alien world, there is too much data—by a factor of several orders of magnitude—for *all* of it to be returned to Earth for display to the humans. Hence the probe must select a *subset* of this environmental data for transmission, and this is one of the key limitations of unmanned space vehicles: the kinds of readings it can best make are those which were anticipated prior to launching. It is easy to imagine situations in which unexpected phenomena, even those so simple that a three-year-old child would recognize them as extraordinary, would be completely missed by the robot probe (e.g., a bird or aerial device soaring above the Viking probe on Mars).

Of course in situations where the environments are totally unknown, it is best to send one-way expendable robot probes to make initial measurements. But when the time comes to make precision measurements, even the best robot probe can be stymied by conditions that its designers did not anticipate (e.g., the inconclusive results of the highly sophisticated Viking life-search experiments or the Viking seismometer which needed only a hard tap with a hammer to fix it, but there was nobody to wield the hammer). And the second major problem surfaces here—*control lag*—since command signals only travel at the speed of light. This is bothersome even as close as the Moon (as the Russians who "drove" robot moon buggies learned). It can be crippling on other planets. Onboard computers can help, but not with the most important problems.

THE COST FACTOR

An argument often heard in favor of unmanned probes is *cost*: they are far cheaper than manned space shots. Extreme caution must be made in evaluating such a claim because valid comparisons must compare like to like. Most "cheap" unmanned probes are indeed the best choice because simple preprogrammed measurements are their forte. They are designed to refrain from doing the kinds of things that people do best. Humans notice unexpected events, react quickly to them, and counteract the all-too-unavoidable mechanical failures of space vehicles.

Let's look at a specific example: the flight to the Moon during the Apollo project. The Russians never sent men to the Moon and claimed that they had never intended to (that's false; they tried hard but were incapable of beating the Apollo timetable); they said that they had rationally chosen the route of "safe, cheap, flexible" moon robots. Some of these robots, the *lunokhods,* drove across the lunar surface for months at a time under the control of drivers on Earth; others (the *scoopers*) loaded samples of lunar soil into small return rockets and blasted them back to Earth. The question is: Which approach was more cost-effective, the Apollo or the Soviet unmanned robot exploratory vehicles?

Consider what Apollo harvested: almost a thousand pounds of carefully catalogued and selected rocks and soil, gathered by trained eyeball-directed hands; tens of thousands of photographs, from close up, from panoramic views, from orbit with high-precision mapping cameras; emplacement of five nuclear-powered stations on the Moon, equipped with highly sophisticated instruments that functioned for up to eight years; active seismic experiments; subsatellites; and other activities. The Soviet results, in contrast, consisted of three grab-bag samples weighing only ounces and two traverses by robot moon buggies, which covered about ten miles and collected television images and some simple physical measurements.

Comparing the most advanced Soviet moon robots to an American flight such as *Apollo 15,* it is clear that probably a dozen robot missions would only begin to approach the pure scientific return of the manned expedition.

What about cost? Each Apollo flight cost approximately four hundred million dollars, and although Soviet propagandists *claim* that their unmanned missions are twenty to a hundred times cheaper than corresponding manned missions, this is pure fantasy. Most space experts compute the cost of an unmanned Moon shot, along the lines of Russia's program, as exceeding one hundred million dollars, probably considerably more.

Now for the arithmetic. To produce the equivalent of one American manned lunar expedition, in scientific information, the Soviets (or a hypothetical American NASA which had chosen to spurn manned spaceflight) would have to launch a fleet of robot space probes costing several times as much. So much for the argument that unmanned space probes are invariably cheaper than manned space probes. In a program as scientifically sophisticated as Apollo, men are cheaper than robots.

However, a good case can be made that simple unmanned surface probes should have been launched eight to ten years earlier than the far more sophisticated manned expeditions; this would have given scientists time to develop more productive research programs for the manned flights when they became possible. The real advantage of such an approach is problematical, and it ignores the nonrational way in which space programs, both manned and unmanned, are funded by governments.

Question: Would such an astronaut-versus-robot analogy extend to Mars? At present it is extremely unpopular to advocate manned flights to Mars (the last government spokesman to do so was Spiro Agnew, and we all know what happened to him). But the unvoiced truth of the matter is that the case for manned exploration of Mars is better today than ever before—and the cost would likely be far less than is generally believed, even by the optimists. The high costs of Apollo, which were associated with the need to develop an entire family of new space vehicles and boosters from scratch, would not be repeated for a mid-1990s' man-to-Mars expedition. Such a flight would, for the most part, make use of modifications to existing space vehicles whose development costs had already been absorbed by other programs (see Chapter 15).

THE HUMAN PRESENCE

Now, what does this argument for remote control imply for the space shuttle? Astronauts, after all, are presumably conducting routine, easily automated tasks, and a case can surely be made for flying such a space vehicle totally on autopilot and from the ground. However, even if all the life-support systems were removed, such a step would not significantly increase the efficiency of the space shuttle; most of the spaceship's design was optimized on operational grounds, which have no connection with the presence or absence of astronauts on board! All the advantages of the shuttle—payload recovery, multiple backup options for space operations platforms for conducting research/industrial activities—would not be enhanced at all by the removal of the crewmembers. On the other hand, experience has shown (particularly on the "routine" Skylab and Salyut expeditions of the 1970s) that the unexpected *can be expected* and that human eyes, ears, and noses, too, tied in to human hands via a powerful data-processing system, are well worth the trouble in life-support systems which their presence requires.

Here's an eye-opening analogy: Why aren't routine commercial airline flights done under unmanned control? There's that old joke about just such a system, in which a soothing tape-recorded voice plays over the cabin intercom, ending with the words: "and of course nothing can go wrong . . . (click) go wrong . . . (click) go wrong. . . ." There's a lot of truth in that joke, which advocates for the elimination of manned spaceflight ignore. Until the day when commercial aircraft routinely fly without pilots, it is absurd to talk of cargo-carrying reusable winged spacecraft flying without pilots.

The most rational approach remains the complementary astronaut-robot strategy, with each option being applied to problems it is best qualified to solve. The lessons of the past decade of manned spaceflight are clear. People have a crucial role to play in space—as pilots, experimenters, repairmen, builders, philosophers, and solvers of the enigmas unveiled by the voyaging, pioneering, exploring, surveying robot vikings. And space travelers are more than just button-pushing meaty-robots. Some of them are going to be the world's best scientists.

PAYLOAD SPECIALISTS

When astronaut candidates visit the NASA space center in Houston for medical and psychological screening, they are issued temporary visitors' badges to allow access to appropriate buildings. But they also carry a more visible sign of their purpose in Houston, a sign which employees have come to recognize with considerable amusement. It is a distinctive blue bag carrying urine samples.

One of the medical tests requires the would-be astronauts to provide hourly samples of urine throughout the day. For this purpose they are given many sample bottles and a case for carrying them as they travel around the center visiting other offices. So whenever employees see people carrying these bright blue bags, they greet them with friendly teasing about the contents of the bag—and with genuine good wishes for success in the screening process.

On occasion, tourists with coincidentally similar handbags have been very confused and embarrassed by the mistaken attention. But usually the identification is accurate.

One group of blue bag carriers was from several scientific laboratories. The men were candidates for the role of payload specialist astronauts on specialized astronomy Spacelab missions in 1986–87. In the spring of 1984, more candidates for full-time astronaut positions, as pilots and *mission specialists*, were carrying their urine bottles around Houston. But the majority of such visitors are now candidates for the job of payload specialist.

The job is essentially a specialist in particular experiments or hardware which would require too much training for a career NASA as-

Spacelab scientists in orbit, on STS-9, showed value of human presence in orbital laboratories, just as humans are productive in earthside laboratories. (Photograph courtesy of NASA.)

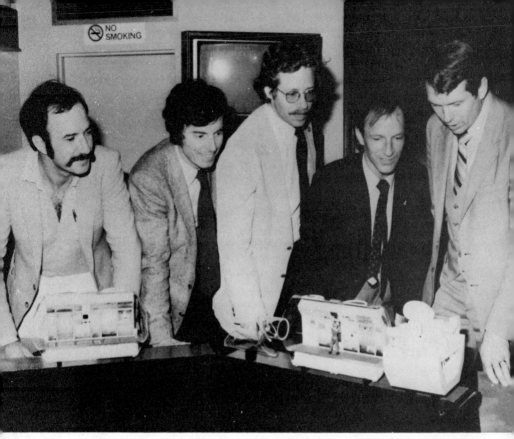

"Visiting scientists" for Spacelab missions include *(left to right):* Wubbo Ockels (Holland; *Spacelab D-1*, late 1985); Ulf Merbold (Germany; *Spacelab 1*, December 1983); Michael Lampton (United States); Claude Nicollier (Switzerland; *Spacelab EOM-1*, mid-1985); and Byron Lichtenberg (United States; *Spacelab 1*, December 1983). (Photograph courtesy of NASA.)

tronaut to obtain. Often, the sponsors of certain equipment have the right to send along one of their own engineers or scientists as part of the space transportation services they have paid for.

The first such payload specialists were aboard *Spacelab 1*. Byron Lichtenberg and Ulf Merbold were career scientists who were experts in the scientific equipment installed aboard the laboratory module. Only in a secondary role were they trained as "real astronauts."

Starting in late 1984, payload specialists from the U.S. Air Force will be going along on missions which carry Defense Department cargo. These men and women are from a group of about two dozen officers called *manned spaceflight engineers*. They are stationed in Los Angeles and work closely with the development of the various Air Force payloads slated to be carried aboard the space shuttle in coming years.

This group forms a pool from which flight personnel can be selected, but it is expected that only a fraction of them will ever actually fly in space. For security reasons their names have not been published.

In early 1985 the *Spacelab 3* mission (*Spacelab 2* slipped a bit more but the numbers stayed the same) will carry two more scientists as payload specialists. Dr. Taylor Wang, a Shanghai-born physicist, is one of them, and his backup scientist is Dr. Eugene Trinh, a naturalized American citizen born in Saigon. The other scientist will be either Loodwiijk van den Berg, a Dutch-born materials processing expert, or Mary Helen Johnston, a scientist from the NASA Marshall Space Center in Huntsville, Alabama.

The training program is still being designed for payload specialists, but it will be very short compared to the years-long preparation of full-time astronauts. Perhaps two weeks of classroom training, along with a month or two worth of reading assignments and a few days of familiarization in space simulators, may be adequate. In the next year or two, a basic curriculum will become standardized.

Payload specialists will be assigned to future Spacelab missions and also as additional crewmembers on routine space cargo missions. While career astronauts fly into space again and again (it is expected that NASA astronauts may make a dozen space missions in their professional lifetimes) the payload specialists will probably fly only once, possibly twice. But many more of them will fly into space: by the end of the decade, the majority of people who have flown into space will have done so as payload specialist astronauts. So many of them will be in screening at any one time that NASA may run out of blue bags. Then we will have to find another way to discover who they are as they wander from building to building in Houston.

4

Spaceflight Geography Lesson

More than a million visitors pass through the NASA Johnson Space Center every year. Wandering around the area, they visit the Museum Building, the cafeteria and gift shop, and several other buildings with public exhibits. They stare at well-known astronauts as they walk past them on the sidewalks, but usually these days nobody recognizes even the astronauts from the most recent space missions. Many of the visitors are well-informed space enthusiasts. Others are merely curious and ask passersby where on the center they can see the space shuttle launching pads. These tourists are lost by more than a thousand kilometers.

Many centuries from now, our space-born descendants may only barely remember the names of the nations involved in the great space breakout of the late twentieth century, far in their past. They certainly will be forgiven for not knowing the actual locations of Earth's first space ports. After all, few Americans today know the home port whence sailed Columbus, or Magellan.

But those of us alive at this moment in history have no excuse for not knowing exactly where the action is. While it may be easy to laugh at the ignorant tourists who look for launchpads in Houston, it is in fact safe to say that even most of the well-informed space enthusiasts would not know the true location from which American manned spaceships are launched.

Those who believe that Cape Canaveral and, for the Soviet side, Baikonur, are the locations of manned space launchings, are in error. For the Soviets, it is a matter of deliberate deception, but for the Americans it is merely one of laziness and habit. The question is not simple at all. Let me present a long overdue lesson in spaceflight geography.

The last American manned spaceflight from the place called Cape Canaveral was the *Mercury 9* flight in May 1963. Six months later, President John Kennedy was assassinated and the cape was renamed Cape Kennedy, in recognition of his role in initiating the Apollo moonflight project. Ten manned Gemini flights and the first manned Apollo flight (*Apollo 7* in October 1968) were launched from this strip of land off the east coast of Florida.

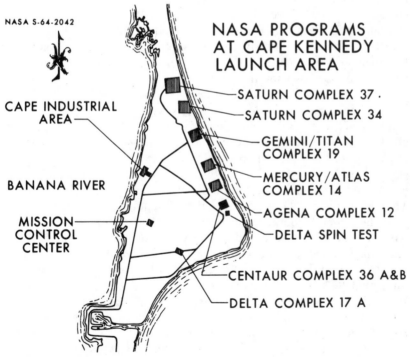

NASA S-64-2042

NASA PROGRAMS AT CAPE KENNEDY LAUNCH AREA

SATURN COMPLEX 37 ·
CAPE INDUSTRIAL AREA
SATURN COMPLEX 34

GEMINI/TITAN COMPLEX 19

BANANA RIVER
MERCURY/ATLAS COMPLEX 14

MISSION CONTROL CENTER
AGENA COMPLEX 12
DELTA SPIN TEST

CENTAUR COMPLEX 36 A&B

DELTA COMPLEX 17 A

Many famous launchpads are located on Cape Kennedy/Canaveral. The last manned launch from here was *Apollo 7* in late 1968. (Photograph courtesy of NASA.)

PAD 39

To launch giant Saturn 5 moon rockets, an entirely new facility was constructed. It was called Pad 39, and *Apollo 8* was the first manned flight from that pad, in December 1968, bound for the Moon. Meanwhile, unmanned launchings of rockets of the type Thor, Atlas, and Titan continued (and to this day, continue) from the Eastern Test Range on Cape Kennedy itself. But Pad 39 was not on Cape Kennedy (which was renamed Cape Canaveral in 1974).

NASA had purchased vast tracts of new land on the Florida coastal islands just north of Cape Kennedy–Canaveral. This new base was named the Kennedy Space Center, or KSC. All of the launch facilities, including the giant Vehicle Assembly Building (VAB) and the shuttle landing runway (built much later) are on Merritt Island, which is separated from the Florida mainland by the Indian River. Except for a tiny neck at its northernmost end, Cape Canaveral is in turn separated from Merritt Island by the Banana River.

And Pad 39 is right in the middle of the oceanside coast of Merritt Island. There are in fact two launchpads there, Pad 39 A and Pad 39 B. Almost all manned launchings have been from A (28° 38′50.9″ north latitude, 80° 38′08.1″ west longitude); B has only been used for *Apollo 10* and the four Saturn 1B missions of Skylab and Apollo-Soyuz. All shuttle launchings are occurring from A, but B will again become operational in May 1986 when two Jupiter-bound spacecraft must be launched aboard two shuttle flights within one week.

But despite the hard facts of accurate geography, the magical phrase "The Cape" is still in wide use among spaceflight workers, the news media, and the general public. Few of them have ever looked at a map of the area in question. And every year, thousands of misled tourists, searching for the NASA moonport's museum and the launchpad tour, get lost in Florida when they drive to the Cape Canaveral road which leads them to the closed-to-the-public unmanned launch areas instead. They have made a detour of more than a hundred kilometers!

This general ignorance has been personally profitable for me. I've won (and no doubt, will continue to win, except where this book is read) many wagers with coworkers with the simple question of where the space shuttle is (or is not) really launched from. Most people still think that "The Cape" means Cape Canaveral (and some newsmen, nostalgic for the Apollo days, still say Cape Kennedy a decade after that name was erased). At least these people think that way until they pay off their wagers. For some, the geography lessons have been expensive, but afterwards they remember about Merritt Island!

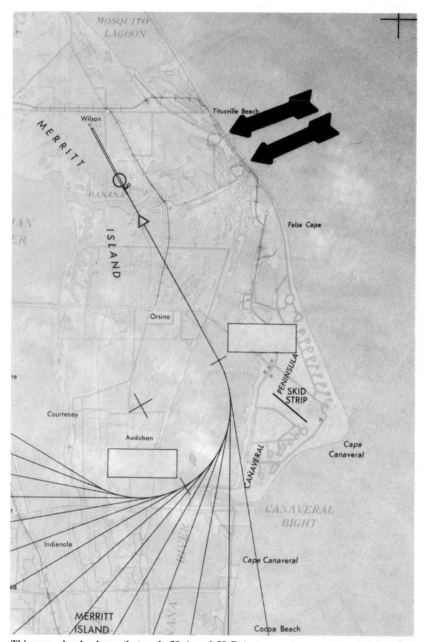

This map clearly shows that pads 39 A and 39 B *(see arrows)* are *not* on what is called Cape Canaveral, but are instead located on nearby Merritt Island. (Map courtesy of NASA.)

Cape Canaveral viewed from Skylab in 1973. Arrows point to Saturn pads 39 A and 39 B, now shuttle pads. (Photograph courtesy of NASA.)

SOVIET GEOGRAPHICAL DECEPTIONS

The Soviets have made cartographical deception into a fine art. And their space launch facilities are an extreme example of this pattern. The busiest one did not even exist officially until 1983; the most famous one is a fiction.

According to official Soviet accounts, their major spaceport is the Baikonur Cosmodrome in Central Asia. Affidavits filed with records certification boards for international flights carry a launch site latitude and longitude near this small village.

But the center is actually located more than two hundred miles away, just north of the town of Tyuratam on the Syr Darya River. In the 1930s, political prisoners (the *zeks* made famous by Solzhenitsyn's chronicles of the "Gulag Archipelago") had dug some stone in open-pit mines there, and a fifteen-mile railway spur had been laid due north from Tyuratam. Twenty years later, chief Russian space engineer Sergey Korolyov (himself a former zek, perhaps a slave laborer at those very same stone pits) had organized the building of rocket launchpads at the edges of the quarries, using the pre-existing deep holes as exhaust flame pits. The rocket range was inaugurated in 1957, and by 1961 Moscow had concocted the "Baikonur" camouflage. By then, however, American U-2 spy planes had made numerous photoreconnaissance sorties which precisely mapped the actual location of the so-called Baikonur Cosmodrome.

The original Soviet launchpad, still in use for manned Soyuz launchings, is located at 45.9235° north latitude, 63.3392° east longitude. Two other pads were later built for the same type of rocket. They are at 45.9983° north latitude, 63.5606° east longitude and at 46.0064° north latitude, 63.5806° east longitude. In addition, there are at least two dozen other launchpads at Tyuratam for different types of rockets. But official Soviet documents still list their cosmonaut launchpad location at 48.2° north latitude, 67.5° east longitude.

Perhaps someday the Soviets can quietly end the deception by renaming the facility as the Korolyov Cosmodrome. It would be a fitting tribute to the man who started it all.

Even more amusing is the history of the Plesetsk Cosmodrome, north of Moscow. Built in 1960 to host ICBMs aimed at America, the center launched its first satellites in 1966. By the 1970s it accounted for an absolute majority of the whole planet's space traffic. Since it launched mostly military payloads, the Soviets never said a word. But Western observers soon pinpointed the location by backtracking the paths of satellites launched from there.

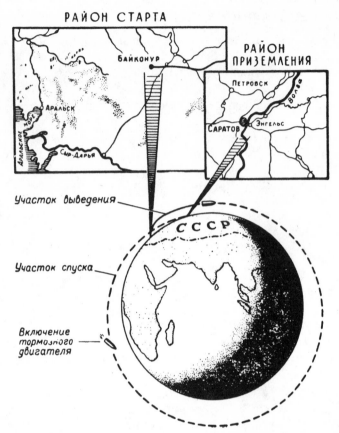

Double Soviet geographical deception appears in this map of Yuri Gagarin's single-orbit space flight of 1961. The path is shown starting from Baikonur *(box on left)*, instead of the true point of origin 200 miles to the southwest. And the space capsule's landing zone, near the Volga River *(box on right)*, is shown east of the launch site, allowing the path to form one complete circuit of Earth. Since the Volga River is nearly a thousand miles *west* of the launch point, Gagarin did not (as the drawing indicates) completely circle the planet.

In mid-1983, *Pravda* officially disclosed the existence of Plesetsk. It was forced to do so by widespread public panic over "flying saucers" (actually only exhaust plumes from rocket launchings). As the Russian public grew near-hysterical over the bizarre apparitions, the Western press had been full of humorous accounts of a disintegrating official coverup. The truth was less painful than public panic.

Personally, I claim the credit. The fact that the spectacular Russian UFOs of 1977–81 were caused by secret military space shots was the conclusion of an exclusive, original investigation I had conducted under the auspices of the "Committee for the Scientific Investigation of Claims of the Paranormal." The resulting publicity made the Soviets look very foolish, as indeed they were. And that embarrassment must have contributed substantially—if not totally—to the deliberate decision to unveil the Plesetsk cosmodrome.

Since mid-1983, no other Soviet magazine or newspaper has mentioned Plesetsk again. Perhaps the admission was a "mistake"; perhaps the UFOs will return. In any case, the reality of Plesetsk has overcome the geographic camouflage draped around it!

WHAT'S IN A NAME?

The names of both U.S. and Russian launch sites have original meanings far removed from the modern connotations. "Canaveral" is Spanish for "canebrake," and the Cape's name can be found on explorers' maps going back to the seventeenth century. It is easy to joke about why the Soviets refused to use the name "Tyuratam"—in Kazakh it means "arrow burial ground." "Baikonur" (actually, "Baikonyr") means "master with light brown hair." The root word "pleset" (with the "-sk" ending implying location) is obscure in old Russian, but it may be connected with a word for the many small islands in the nearby river. And "Kapustin Yar," the smallest and oldest Soviet rocket range (on the lower Volga River), means "cabbage crag."

SPACE COMMUNICATIONS

Another aspect of geography is distance: What is the shortest path between two points. And what are the paths along which communication can flow? It turns out that the shortest paths are rarely the most useful ones! Let me illustrate with an example.

On the overhead television monitors throughout Mission Control, the countdown to "A-O-S" (Acquisition of Signal) neared zero. People drifted back toward their duty stations. Men and women glanced up at the flashing numbers and sat straighter at their consoles. Cans of soft drinks and half-eaten sandwiches and doughnuts were pushed aside. Conversations ceased. Everybody adjusted their miniature communications headsets and made sure that all the "air-to-ground" voice signal buttons were illuminated on their consoles—if not, they pushed the buttons with a quick jab of a finger, activating the channels. Visitors and off-duty associates stepped back from the consoles of the on-duty flight controllers, so as not to offer distraction.

The digital clock reached zero. A hundred pair of eyes scanned panels of warning lights, alert to see if any flashed red the moment the fresh telemetry data flooded down from orbit. In a hundred pairs of ears in Houston, and in a few pairs in space, the crisp voice of the CAPCOM sounded forcefully. "Columbia, Houston," called the astronaut on communications duty at Mission Control. "Acquisition of Signal—A-O-S—through Utopia for six minutes. How do you read me?"

For the next several minutes there was a brief opportunity to discuss the progress of the flight with the astronauts. How far were they along in their checklists? Were there any new problems they needed advice on? Changes had been made to the upcoming activities schedules and the crew had to be told.

Flight controllers noted the condition of equipment on the spaceship and the quantities of fuel supplies ("Any signs of unexpectedly high usage?" they asked themselves). Detailed entries were made into logbooks. On television screens, columns of numerical and graphical data appeared, and at the push of a button, paper copies of the displays were dispatched to the consoles through pneumatic message tubes.

All too quickly, the CAPCOM astronaut reported, "Loss of Signal—L-O-S—in twenty seconds. See you at Atlantis in thirty-one minutes." The rush of activity relaxed as engineers evaluated the latest readings, projected what conditions should be on a healthy spaceship at the next communications pass, and prepared to observe playbacks of data dumped at high speed during the last pass. And for a few moments, there was time to relax, to consult colleagues, to attend to personal needs.

This has been the normal rhythm of our work in Mission Control for more than ten years. But it has begun to change. The brief interludes of live communication, interspersed between long periods of silence, are being replaced by nearly continuous contact. The key is a satellite called TDRS—Tracking and Data Relay Satellite.

The Tracking and Data Relay Satellite

We knew the transition would be difficult since it would mark a revolution in space communications. But nobody really anticipated the plague of booster problems that nearly sank the first TDRS in April 1983 and pushed off any new launches for at least a year. And even after the errant *TDRS 1* finally limped bravely up to its proper orbit 24,800 miles high early in July 1983, more troubles awaited it in the form of ground processing computer software flaws. Still, the effort will eventually work—and it is worth it.

Communications with moon-bound Apollo spaceships in 1968–72 were continuous, not sporadic like Skylab or space shuttle. The spaceships were so high that a single ground station could remain in touch for hour after hour as the Earth slowly turned.

Not many of today's flight controllers remember that time, however. Only perhaps 20 or 30 percent at most of the men and women in Mission Control for space shuttle flights are old enough to have been at work on Apollo. Most of them were accustomed only to the staccato pattern of AOS and LOS.

Today's spaceships in near-Earth orbits face the paradox of poorer communications, brought on, strangely enough, because they are so much closer to the ground tracking sites. The reason is this: The rapid motion of the satellite a few hundred kilometers up means that it will "rise" and "set" over any geographic point in a very short interval.

One radio station can listen in for only eight minutes at most. And the next time the satellite comes around, 1½ hours later, Earth's rotation may have carried that station too far to the east for any reception at all.

Such periods of loss-of-signal are not merely inconvenient; they can be dangerous. The spaceship's multitude of mechanical and electrical systems are not being closely monitored for misbehavior. True, alarms will sound on board if things get bad, but flight controllers can often detect early trends and then advise action to forestall trouble— but only if they are watching live data. And if something catastrophic occurs, such as the explosion aboard *Apollo 13* in 1970, Mission Control's rapid and well-informed advice can make the difference between life and death.

The complete TDRS system will consist of two active satellites and a spare, talking directly to a ground control station at White Sands, New Mexico. The TDRS West satellite will be in a 24-hour orbit hanging perpetually over the equator at 171 degrees west longitude, or 64 degrees west of White Sands. The TDRS East satellite will be over the equator at 41 degrees west longitude, or 66 degrees east of White Sands.

With this system, coverage for satellites close to Earth is almost complete, but not quite. There is a blind zone over the globe opposite White Sands. This forms a north-south strip which runs from Central Asia south across India and deep into the south-central Indian Ocean. And transmissions from Soviet air defense radar systems may also interfere with satellite communications over or near the USSR.

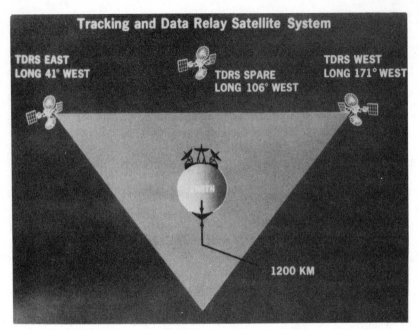

For the Tracking and Data Relay Satellite system, two satellites (plus a spare) will provide constant communications except over a narrow north-south strip over the Indian Ocean and Central Asia. (Photograph courtesy of NASA.)

But in general the improvement will be nothing short of revolutionary. Instead of the 20 percent coverage in a typical orbit (say, eighteen to twenty minutes, from two or three ground sites), there will be 85 percent coverage or better. For perhaps twenty minutes per 1½ hour orbit, the astronauts will be out of contact with Earth.

The tremendous advance in convenience, economy, and safety is undeniable. It will also allow a much greater volume of scientific data to be returned from space vehicles, both manned and unmanned. But in return, the engineers of Mission Control will have to adjust to the entirely new pattern of operation.

Space Fleets

A quotation from President Calvin Coolidge used to be displayed on a bulletin board in the astronaut office at the Johnson Space Center in Houston. In the 1920s, the parsimonious American chief of state was exasperated by a budget request for more airplanes for the fledgling Army Air Corps. Coolidge did not understand the need for so many. "Why don't the aviators just buy one plane," he asked querulously, "and then take turns flying it?"

The subtle humor of the quotation was connected with the fact that up until early 1983 the U.S. Space Transportation System was just a glorified "Calvin Coolidge Spaceline"—it just had one spaceship, and the astronauts took turns flying it. But then, on April 4, 1983, the *Challenger* roared into space and became America's second space shuttle vehicle, giving the United States the beginnings of a real fleet of spaceships.

In 1983, NASA's plans were to build two additional space shuttles—called *Discovery* and *Atlantis*—and spare parts for a fifth, if needed. This foursome will be the mainstay of American space traffic until at

Building a space shuttle: *Discovery* (Orbiter #103) at the Palmdale, California, plant of Rockwell International. (Photograph courtesy of NASA.)

least the early 1990s. But all four will not become operational until late 1985.

Discovery appeared in space in 1984. Together with *Challenger,* it carried routine space payloads all through the year. *Columbia* received one flight and then was removed from service for a year to undergo major structural modifications.

In mid-1985, *Atlantis* is to make its orbital debut. Shortly before that event, *Discovery* will be transported to the new launch site at Vandenberg AFB, California, leaving *Atlantis* and *Challenger* to carry space traffic until the end of the year. Then, the rebuilt *Columbia* will return to service while the *Discovery* will make an inaugural flight into polar orbit. At that moment, the four-ship fleet will become a reality.

The current "payload manifest" for 1986 launchings can serve as a sample of typical space traffic in the late 1980s. Seventeen launchings are planned: five each for *Columbia, Challenger,* and *Atlantis,* and two polar orbit flights for *Discovery.* Four of the flights will carry military space payloads (about 24 percent); about a quarter of the cargo will be scientific; the remainder of the missions will be devoted to applications such as communications and weather satellites. Exact cargoes may well change, but this mix is typical for the period of mature space shuttle operations in the late 1980s.

The space fleet will have become operational. A true sign of the coming-of-age of the space shuttle will be the first time that a manned launching is not mentioned on the American national evening television news. The sooner that occurs, the better—so that spaceflight can move from the extraordinary phase into the everyday phase of operations. That time is not far off.

IMPROVING THE SHUTTLE

Every transportation system can be improved, once it has worked the first time. That engineering maxim applies to the NASA space shuttle, too, which is the subject of new analysis efforts to determine how it can form the basis for a whole new generation of space launch vehicles for utilization well into the next century.

Several avenues of improvement are open for NASA engineers. First, the current shuttle engine systems can be uprated to power levels higher than first designed. Second, modifications and replacements can be made into the "stack" of winged Orbiter, giant External Tank (the original "ET"), and the twin Solid Rocket Boosters (two of which still lie at the bottom of the Atlantic). Third, components of the space shuttle stack can be rearranged with new or highly modified equipment to produce booster systems with much greater economy, capability, or flexibility—but without the expense of a new development program since the equipment has already been proved out during regular shuttle missions.

Upgrading the Engine System

The first approach is an evolutionary one and it depends on capabilities which had been designed in from the start as margin for unpredictable weaknesses in the total system. In one case, the engines were supposed to operate at 100 percent rated thrust, but to guarantee that level they were overdesigned to provide at least 107 percent thrust and probably quite a bit more. Other systems were designed for "worst cases" which, once the *Columbia* actually flew, were shown to have been unnecessary—in hindsight, that is; several of the systems which were built for worst case stresses turned out to have been constructed

The giant External Tanks are used only once and then thrown away (it would cost more than they are worth to recover them). Someday they may be collected in a junkyard in orbit where their scrap metal can be recycled. (Photograph courtesy of NASA.)

just strong enough. The prudence paid off since the whole system worked. And now that same prudence can be cashed in regarding other systems which did not really need to be built as powerfully as they were. That margin can in many cases be converted into a greater overall capability without affecting the safety or reliability of the system, now that the actual (rather than feared) operating conditions have been charted.

An example is in the area of the Space Shuttle Main Engines. Three engines, fuelled with liquid hydrogen, provide a nominal 470,000 pounds of thrust, which is considered 100 percent. There was enough margin left in the propellant tank for several extra seconds of burn time, above and beyond the 510 seconds needed to achieve initial parking orbit. Now that many flights have been made, the actual propellant consumption rate has been determined and the uncertainties reduced. The result is that for the same prelaunch propellant loading in the External Tank, NASA engineers think they can squeeze several additional seconds of SSME thrust and still maintain a safe margin against the tank running dry prematurely. Those several extra seconds convert to more than one hundred miles of additional altitude in the parking orbit, a bonus which thus need not come out of the smaller onboard maneuvering engine system and hence can be directly converted to either a heavier payload or a higher operating altitude.

While the SSMEs on early flights was only guaranteed for 100 percent thrust under nominal conditions and 107 percent in emergencies (such as if one of the three engines failed), testing was completed on SSMEs which could be rated at 109 percent for nominal lift-off. Few hardware modifications are required to squeeze this extra power out of the engines. The major testing, at NASA's National Space Technology Laboratories near Bay St. Louis, Mississippi, is designed to develop the required confidence in such an improvement.

Hardware Modifications and Replacements

Although upgrading the SSMEs to 115 percent or even 130 percent would require some hardware modifications, it is probably feasible. According to Dominick Sanchini, a Rocketdyne Corporation official and manager of its main engine program, "We think we have a firm basis for saying this engine does have the potential for growth and certainly to a power level of 115 percent of normal thrust. This would probably be possible with no significant design changes. With more significant changes we would have the capability to go up to 130 percent." The greater thrust would be converted to greater payload delivered into orbit, particularly on the more difficult polar launches from California in the mid-1980s.

Similar evolutionary improvements include lightweight external tanks and solid rocket boosters. Such pounds saved in structural weight convert directly both into payload on orbit and into safer emergency abort procedures. The first lightweight ET was used on shuttle mission seven and provided about 3,000 pounds of extra payload. The lightweight SRBs are farther in the future.

A long-planned but still unfunded SRB improvement involves what are called *filament-wound casings*. Most of the metal segments of the current SRBs would be replaced with graphite filament material. This would reduce the empty weight of the SRBs by one-third and would translate to a performance increase of 6,000 additional pounds of payload into orbit. Since such an advantage would accrue primarily to polar orbit launchings, there is still dispute in Washington over whether the U.S. Air Force, the primary customer for such launchings, should be directed to pay for the development program for the filament-wound SRBs.

New Booster Systems

The next logical increment in upgrading the space shuttle involves the addition of unique hardware. One approach concentrates merely on increasing payload-carrying capacity. Another proposal is a solution of current restrictions on the width of shuttle-carried payloads (they must fit into the fifteen foot diameter payload bay).

The most underpowered portion of the space shuttle ascent is after the two SRBs burn out and fall away, about two minutes into the flight. The huge ET still contains three quarters of its original propellant load, but the vehicle thrust has been reduced from six million pounds to about 1½ million pounds. Within a few minutes, however, as the propellant load is burned off at three thousand pounds per second, the vehicle's thrust-to-weight ratio improves significantly.

To provide more push at this critical phase, engineers have designed a *boost module* based on proven components of the standard Titan missile. The unit would be installed under the shuttle's External Tank and would be ignited a few seconds after the shuttle had cleared the launchpad. It then would burn for a little more than three minutes. Since the unit would initially weigh 385,000 pounds and would develop 529,000 pounds of thrust, its contribution to the early phases of the ascent would be minimal (less than 2 percent improvement). Where the unit would really pay off, however, would be right after the SRBs were dropped when the shuttle was riding on its three SSMEs alone. At that point the proposed unit would provide about 400,000 pounds of excess thrust (almost the equivalent of a fourth SSME) for another 1½ minutes. At burnout this boost module would be jettisoned and not recovered.

Performance improvements would be substantial. At current power levels, the space shuttle can carry about 24,000 pounds into polar orbit. With the boost module, that figure could almost be doubled, to 41,000 pounds.

The space at the end of the ET can be utilized for other purposes as well. For payloads with diameters larger than the diameter of the Orbiter's payload bay, a special cargo cannister can be built and bolted in place below the ET. Objects up to 27 feet in diameter (such as giant mirrors) could be carried inside. The cargo cannister would have to be grabbed by the Orbiter's Remote Manipulator System (the Canadian RMS, or "arm") before the ET plunged back into the atmosphere and burned up. Such an improvement would not carry heavier payloads into orbit, merely wider ones.

Shuttle-Derived Vehicles
Liquid Rocket Boosters

Plans to upgrade the SRBs may eventually give way to proposals to replace the units entirely. One design would involve Liquid Rocket Boosters (LRBs) each carrying five SSMEs to generate the required thrust at lift-off. Since recovery of such delicate equipment from the ocean would present severe problems, it has been suggested that the units be equipped with wings for direct flight back to the Kennedy Space Center launch site.

Although the development of the LRB concept would raise the shuttle payload capacity to 100,000 pounds into low orbit, detailed traffic analyses have failed to show any cost advantage. One study by Martin Marietta Aerospace Corporation revealed that both operational costs and volume limitations of the Orbiter bay would vitiate any performance advantage of such a system unless its fly-back capability were extremely reliable.

The Big Dumb Cargo Module

So for significantly heavier payloads, the evolutionary improvement strategy must be abandoned. The leading candidate for such a step involves dispensing with the winged Orbiter (and the astronauts!) entirely and replacing it with a Big Dumb Cargo Module.

Such a configuration would preserve the "side-mount" posture of the Orbiter relative to the ET and SRBs. In place of the winged Orbiter would be a long cylinder quite similar in dimensions to the Orbiter's fuselage. At the aft end of the cargo module would be the three SSMEs and associated orbital maneuvering and altitude control rockets—the same type of hardware mounted on the aft end of the Orbiter. But in this case they would be grouped in an easily detachable module which would either return from orbit on its own or be picked up (after dismantling) by a subsequent Orbiter mission. The cargo module could carry up to 130,000 pounds of payload, twice the standard load of the baseline space shuttle, and payloads of medium width could readily be accommodated.

In exchange for this doubling of payload capability, this design would give up the option of returning a payload to Earth, either routinely or in the event of malfunction. Also, the presence of human operators would be sacrificed. However, traffic analysis has shown that this bargain would be acceptable to many classes of payloads

The most likely form of a Shuttle-Derived Vehicle (SDV) could look like this and have the payload capacity of an obsolete Saturn 5 at a fraction of the cost. Unmanned cargo carriers such as this will be needed to assemble the spaceships for Mars. (Photograph courtesy of NASA.)

already projected for space shuttle manifests of the early 1990s. Additionally, any ambitious new space projects which involved major construction activities or the assembly and fuelling of space vehicles aimed farther out into space would find such a system highly desirable. The Martin Marietta study, confirmed in less rigorous fashion by other analyses, points to substantial operational savings following the development of such a booster combination.

The In-Line Variant

An alternate architecture for such a heavy-capacity Shuttle-Derived Vehicle (SDV) is the "in-line" variant. Here, the propulsion module is mounted directly below the ET and the cargo module is mounted directly atop it. This system looks more like a conventional rocket but might require more substantial modifications to standard shuttle hardware. On the other hand, one option utilizing four SRBs might be able to carry in excess of 200,000 pounds into orbit—the equivalent of the now-obsolete Saturn 5 of Apollo and Skylab glory.

The SRB-X

Shuttle-Derived Vehicles can also be developed in the other direction—for smaller payloads than currently envisaged for mainline space shuttle missions. Users might require a launch which does not utilize the full capability of the standard system; yet they may not be able to find other payloads willing to share the ride. For such payloads, boosters based on SRBs alone or in combination with other small stages may be worthwhile.

This is the SRB-X concept developed at NASA's Marshall Space Center in Huntsville, Alabama, where the American space agency's primary space propulsion work is conducted. An SRB-X may consist of a single SRB (recoverable!) with smaller booster stages on top, capable of carrying up to 30,000 pounds into orbit, or it could require a triple SRB combination in which two outrigger SRBs carry the middle unit to the edge of space, whence it fires and carries a 60,000-pound payload all the way into orbit. Certain operational advantages could accrue to such booster systems.

The continuing NASA budget crunch has prevented the official approval of any of these alternative Shuttle-Derived Vehicles, but their evident economy and flexibility make them excellent candidates for space launches by the end of the 1980s. By then, space engineers will be well along with designs for new manned space shuttle systems for operational use at the turn of the century. But meanwhile, the powerful Columbia-class space shuttle system has untapped capabilities which may take decades to fully appreciate.

THE COMING OF THE SUPERBOOSTERS

For the Soviets, the 1980s is also going to be a time of transition for space transportation. The family of booster rockets which has served for more than twenty years is going to be replaced at last.

An epochal cosmic event went unheralded sometime in 1982. In a manner of speaking, it was the millennium. Technically, it was merely the one-thousandth launching of the primary Soviet space booster.

The mainstay of the Russian space program for a quarter of a century, the booster was known to its original designers as the R-7 (Rocket 7) or, affectionately, as "Ol' Mark Seven"—in Russian, *semyorka*. Western observers call it the A-class vehicle. In any language it is *the* workhorse of spaceflight. The launching was a fitting tribute to the twenty-fifth anniversary of the first space satellite, *Sputnik 1,* which rode a *semyorka* into orbit on October 4, 1957.

In the past quarter-century, the *semyorka* family carried 6,000 tons of payload into orbit, many times the total weight of the cargo any other type of booster has lifted. If all the rocket stages used in these missions were laid end to end, they would stretch 70 miles—nearly far enough to reach into space. One-half million tons of liquid propellants, mainly kerosene and liquid oxygen, were consumed by the booster.

The *semyorka* has seen glory and tragedy. It carried the first artificial satellite, the first Moon shots, the first men (and a woman) into orbit. It is the only Soviet space booster ever to have been "man-rated" safe to carry crewmen, and it has done so for more than sixty rides. It also once killed dozens of people: a balky model exploded on the launchpad in 1960 during an inspection that violated safety standards the Russians had foolishly set aside to meet a politically inspired deadline. In 1975 another *semyorka* was responsible for spaceflight history's first manned-launch abort when an upper stage malfunctioned and dumped two cosmonauts onto a night-shrouded, snowy mountainside uncomfortably close to the Chinese border. Another *semyorka* blew up on the launchpad in 1983 and two cosmonauts narrowly escaped by using their "launch escape tower."

Even in the early 1980s more than half of all Soviet space shots have been atop *semyorka*-class boosters. Although that might be interpreted as a lack of real progress in Soviet space propulsion, in fact it is probably testimony to the foresight and advanced thinking of the missile's creator, Sergey Korolyov. He persuaded the Kremlin leadership to finance an intercontinental ballistic missile (ICBM) program in 1955, but the missile built was clearly designed as a space booster. Although it never made a very efficient weapons carrier (NATO called it the SS-6) and was quickly replaced by newer ICBM designs, it has held its place as a space-transportation system without equal in the world.

A million pounds of flaming thrust pushes the *semyorka* off its concrete pad, built on a porch overhanging a deep flame pit. The twenty rocket engines are divided into five clusters of four each. The clusters are attached to a long central core unit and four tapering strap-on stages. After two minutes, the strap-on units fall away and the core central stage carries on into space. An upper stage then pushes the payload into orbit.

By 1984, that scenario had been conducted successfully more than 950 times. Several dozen more launches have fallen back from space

The first space satellite booster in 1957, the *semyorka* (Rocket 7) with upper stages which nearly doubled its length, is still the workhorse of the Soviet space program. (Bottom photograph courtesy of NASA and USSR Academy of Sciences.)

failures. Another two dozen *semyorkas* were expended between 1957 and 1959 during a brief weapons-test program.

But now the reign of the *semyorka* is apparently drawing to a close. Several trends are converging toward its extinction. Many standard payloads previously carried aboard *semyorkas* are now being launched aboard more modern SS-9 ICBMs, which have in turn been replaced for military use in the ICBM silos by the even more awesome SS-18 missiles. One bigger booster, the Proton, has, after more than a decade of parallel service, apparently neared the stage where it, too, can be trusted to carry cosmonauts. A special minishuttle is evidently being built to transport cosmonauts to and from a planned large space station later in the eighties. That station itself is reportedly to be launched by

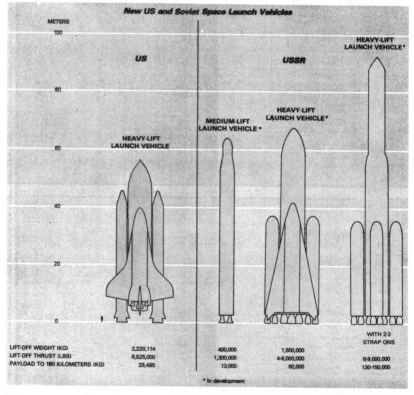

Pentagon experts describe three new Soviet space launch vehicles which share many common subsystems: an intermediate booster to replace the Proton, a giant superbooster to dwarf America's Saturn 5, and a space shuttle remarkably similar to the American design. (Illustration courtesy of Department of Defense.)

a carrier rocket at least twelve times the size and power of the *semyorka*. With these developments on the horizon, the next ten years may witness the launching of the very last of Korolyov's progeny.

The most spectacular successor is bound to be the *superbooster*. Its advent had long been expected by Western observers, and rumors of an impending launching have been circulating for several years. In 1981 the Pentagon officially confirmed that such a booster, with "six to seven times the payload capacity of the space shuttle," was being built. That comes out to about 200 tons in orbit, twice the payload weight lofted by the now-defunct Saturn 5.

Though no official word was given on possible schedules, most experts believe that a first test flight (probably with a dummy payload) could occur by 1985. The booster could become operational by 1988, carrying giant twelve-man space stations into low-Earth orbit as well as Salyut-class three-man craft into lunar orbit, and eventually, on to interplanetary trajectories.

The development of this two-part program, the big booster and the minishuttle, certainly will hold center stage in the Soviet space program of the late 1980s. But every year there will also be dozens of launchings of small (five tons or so) unmanned satellites to continue such current applications as spying, weather surveys, and communications, and even, from time to time, genuine scientific satellites. These payloads will be carried by the last of the *semyorkas,* and when the supply runs out—perhaps the factory has already been shut down and the warehouse is emptying fast—the newer SS-9–type boosters will take up the slack and carry the space cargo.

It is too much to expect that Moscow will officially announce the end of the *semyorka* program. But it can be fervently hoped that when the last of its kind is hoisted up onto the launchpad, a few history-minded engineers will be there to hoist a toast to Ol' Mark Seven.

6

Space Maladies

The last time a team of Russian cosmonauts returned to Earth from a months-long orbital expedition, they were placed on stretchers while anxious doctors monitored their heartbeats and blood pressures. Even after the relatively short week-long missions of the space shuttle *Columbia* in 1982–83, some astronauts admitted to feelings of nausea and weakness. At a postflight press conference, one admitted sheepishly to having vomited—but "only once," he added defensively.

THE PROBLEM OF WEIGHTLESSNESS

Clearly, some aspects of the outer space environment do not agree with the physical well-being of space travelers. Indeed, by now there is sufficient evidence for the U.S. Surgeon General to order the placement of a placard over the entry hatch of every space-bound vehicle: Warning! Space conditions have been determined to be hazardous to your health. If astronauts had to be provided with working conditions guaranteed to be as safe as those mandated for earthside workers, the

Occupational Safety and Health Administration (OSHA) probably would ban manned spaceflight altogether.

These physical effects range from the trivially obvious to the invisibly insidious, from the immediate to the imperceptibly gradual. On the outside, the actual shape of the space traveler's body and face changes noticeably. On the inside, blood and bones and muscles react more slowly and more subtly.

The result is a threat to human health which parallels in many ways the decay of old age: bones grow brittle, muscles grow weak, digestion becomes questionable, balance mechanisms atrophy, the blood loses much of its oxygen-carrying capacity. But instead of taking decades, all of this happens on spaceflights in months.

Is there any way around these problems? While short-term goals are merely to allow crewmembers to stay sufficiently healthy to complete their missions and return to earth, long-term requirements include questions of permanent adaptation to space conditions; that is, what would happen to the physique of a true "space colonist" who intended to live out the rest of his/her natural life in the weightless conditions of outer space?

Such fears are not new to doctors dealing with human reactions to space conditions. A generation ago, chimerical dangers seemed to loom on the horizon of manned spaceflight. Could a body even survive such conditions for minutes, hours, or days? Would hearts burst from overloads? Would spacefarers drown in their own vomit? Was sleep possible under free-fall conditions that should yank a man to wakefulness repeatedly due to his instinctive fear of falling—and lead to quick psychological breakdown? Were there uncharted belts of unknown radiations waiting to sterilize, blind, or quickly fry careless explorers?

These problems did not materialize. People have lived happily and healthfully in space for six or seven months at a time. But all the same, newer problems have arisen to replace the old bugaboos. And there have been some unexpected unpleasant physiological problems, too.

Astronaut-physician Joseph Kerwin, a member of the first Skylab space station crew in 1973, described one such effect. It was connected with general functioning of the human digestive system, the gastrointestinal (or GI) tract. "The only abnormal finding is that it is very difficult to belch," Kerwin noted. This seemingly innocuous problem had a profound influence on another bodily function which in turn made its mark on the air quality index of the space station's atmosphere. "One does swallow a great deal of gas," Dr. Kerwin continued delicately, "and one finds that the GI system processes it downwards very effectively with great volume and frequency." Tight quarters in space aren't bad enough—the astronauts must face orbital flatulence!

Even to casual viewers of space-to-ground televised scenes the unique posture of astronauts in weightlessness is very noticeable. An astronaut "tends to assume a peculiar posture," Kerwin related. "When he is relaxed, his neck extends and moves backwards [much like that of an alarmed ET!], his shoulders and elbows flex twenty or thirty degrees. The same is true of his hips and knees, and the shoulders tend to rise up toward his ears." On Skylab, equipment such as chairs, cabin consoles, and close-fitting spacesuits, all designed for one-G posture and body shape, were found to be awkward and ill-fitting in flight.

The body's actual shape also changes. The waistline reduces up to four inches and the curve of the backbone straightens, increasing one's height up to two inches. This happens because the intestines float upward toward the chest cavity, and gravity's crush on erect posture is removed from the spine.

So in space, people get taller, thinner, and lighter all at once! While these changes might seem desirable to short or obese people, they have a price. The fluid shifts cause legs to get starkly thin—the astronauts called this the "chicken legs of space" effect—and the muscles soon seem lax and doughy.

Physical changes to the astronauts' heads are also very evident. Their faces get puffy and their eyes take on an almost Oriental appearance. Crewmembers experience a feeling of fullness in their heads, much like nasal congestion, while the blood vessels in their necks and scalps bulge uncomfortably.

All of these changes occur almost immediately as bodily fluids concentrate in the upper body instead of being pulled by gravity into the lower parts of a space traveler's body. But this immediate result of spaceflight is itself the cause of more far-reaching effects.

Feedback mechanisms in the brain sense this fluid shift as an excess of body fluids, such as blood plasma and associated red blood cells. A process is thus initiated to purge this phantom excess by reducing fluid intake (crewmembers just aren't as thirsty as they would be on Earth, even though they really need to continue drinking fluids) and by cutting off red blood cell production until a "normal" volume is restored. This process takes several weeks and is one of the subtler effects of weightlessness.

But the rush of fluids to the head may also contribute almost immediately to one of the best-known and the most visible medical problem in space: "space sickness." (NASA prefers the euphemism, space adaptation syndrome—SAS.)

SPACE ADAPTATION SYNDROME

During the first few days of an orbital mission, about one-third of all crewmembers get sick and vomit. Others experience less severe symptoms, such as nausea, dizziness, sensations of rotation and false posture, and loss of appetite leading to fatigue and dehydration. If one stays up long enough, this will pass—but all slated American manned space missions of the 1980s, together with probably half of the Soviet ones, are not scheduled to last much more than a week. Thus techniques are being sought to counteract this problem or to screen candidate crewmembers for those potentially susceptible, an approach which has so far been fruitless because no correlation has been found between people who get space sick in orbit and those with a tendency to get motion sick back on Earth.

"No satisfactory physiologic model exists to explain space sickness," admitted astronaut-physician Anna Fisher in a special report coauthored with her fellow-astronaut husband Dr. William Fisher. "Such sickness may be fundamentally different from the various forms of motion sickness experienced on earth," they suggested, adding that "anti-motion-sickness medications have been of limited value in treating this problem." The best drug, a combination of scopolamine (for the nausea) and Dexedrine (to combat the depressant effects of the first drug) called Scopo-Dex, used in either pill or skin patch form, has, according to the astronauts, "been found to be only partially effective." And as sufferers of seasickness are well aware, once real nausea has set in it's too late to take a pill; it would remain in the stomach and not be absorbed where it was most needed. Insertion in the form of a rectal suppository has been summarily rejected!

What Causes Space Sickness?

Since these "pill regimes" have not worked, space doctors are seeking a better understanding of the original stimulus for space sickness instead of just trying to fight its symptoms. Among American space doctors, the classic explanation for space sickness is called the sensory conflict theory, which claims that "it's all in the head." The theory goes like this. During training on Earth, crewmembers subconsciously learn which way should be up and down in their space vehicles.

However, once in orbit, the astronaut's balance mechanism in the inner ear gives the brain conflicting and confusing reports on true up-down orientation. The brain then expresses its bafflement by inducing vertigo and a form of motion sickness.

If this theory is accurate, one procedural countermeasure is to avoid confusing the brain with wild movements. Crewmembers have been advised to minimize head movements and to keep their bodies positioned close to the cabin's architectural up-down frame of reference for a few days. And indeed this seems to work.

"People who don't wiggle their heads have less tendency to get ill," noted Dr. Joseph Sharp of NASA's Ames Research Center near San Francisco. Doctors have also noticed that the larger space vehicles have grown, the more astronauts move around and the more they experience motion sickness. (That could also be a reporting problem, one cynic suggested: later astronauts are merely more willing to discuss such symptoms with doctors.)

Since this theory involves the functioning of the brain, Sharp and his colleagues believe that teaching the brain new thought patterns could control its vulnerability to space sickness. At Ames, Dr. Patricia Cowings has developed a version of biofeedback which provides symptomatic relief of motion sickness through use of calm-inducing mental disciplines taught to test subjects. Presumably such techniques would also be effective against space sickness.

"We have some evidence that biofeedback is as good as or better than the drug approach," Sharp said recently. He pointed to "very suggestive" results at the Air Force's Brooks Medical Center in San Antonio, where aircraft crewmembers were successfully taught to overcome otherwise-disqualifying tendencies towards airsickness.

A training program consisting of eight to ten hour-long sessions might be all that is required, according to Ames doctors. But other medical experts, particularly at the astronaut center in Houston, remain skeptical. One even derisively dismissed the technique as "that trick about spooking yourself." Others suggested that the training program was grossly underestimated and that it could take a hundred hours or more to reach a useful competence; this requirement would be out of the question for packed training schedules.

For their part, Soviet space doctors have expressed little interest either. They prefer to seek a direct physiological basis for space sickness and are now concentrating on what they call the peripheral endolinth problem. In this theory, the fluid shift to the head is seen as the direct culprit since it causes a pressure change across a membrane in the inner ear which in turn sends spurious signals out along the receptor side of the eighth nerve. And indeed there is some conflicting flight data that does suggest that the vestibular function is not working properly. If this is really the problem, a physical or mechanical control of the body's fluid shift might provide relief, and various devices are now being tested on short "visiting missions" to Soviet space stations. Otherwise, an alternative technique might be some sort of preflight conditioning of crewmembers.

And there are some even more far-out theories. According to Sharp at Ames, "The sickness may be set off by a peculiar set of neurons ejecting proteins into the cerebral spinal fluid on the third ventrical." This could lead to a metaphorical short circuit (much like the reflexive sneeze caused by the sudden presence of a bright light). Sharp went on: "Sitting on the fourth ventrical is the vomit center in the brain," and that might get triggered unintentionally. If this theory is correct, a countermeasure would involve trying to identify that protein and then finding a drug to act as a blocking agent. The crewmembers could then just take a pill or an injection shortly before launch.

Long-Term Adaptation

Whatever the cause of space sickness, it fades within a few days. Spacefarers who remain in orbit longer soon recover, only to face a different class of space maladies.

This type of problem involves the long-term adaptation of the human organism to space conditions. This adaptation does not threaten to detract from mission performance as long as the space travelers remain in orbit, but it can seriously threaten their health once they return to the full-weight conditions on Earth.

Familiar television images of exercising astronauts have been sent back to Earth for more than a decade. The logic for such strenuous and time-consuming activity is straightforward: without their stresses of everyday weight, the human body's muscles and heart/lung system ease off and "get lazy" to the point where the stresses of a return to Earth could be traumatic, even life-threatening.

Soviet cosmonauts have gained valuable experience in this problem during repeated half-year orbital excursions. Their onboard mini-gymnasium has included a treadmill, spring-loaded exercisers, a bicycle, plus special elastic trousers which pull the legs up against the body unless a muscular counterforce is constantly exerted (this device was nicknamed the Penguin Suit because people on Earth tended to waddle around penguinlike while walking with it on). In all, cosmonauts visiting *Salyut 6* in 1978–81 and *Salyut 7* in 1982–83 daily went through a two-hour exercise regimen which was the work equivalent of climbing the stairs of a two hundred–story skyscraper. On the longest flights of the series, this practically added up to climbing on foot into space.

With such preparation, the Russian space voyagers have been able to walk immediately after landing (although ever-anxious doctors make them lie down whenever possible). The day after his return from a six-month flight in 1979, Valeriy Ryumin walked half a mile through a park outside his hotel, and by the third day he was jogging for ten minutes at a stretch. His readaptation to Earth conditions was rapid and complete, so much so that eight months after his landing Ryumin was allowed to blast off again for another six months in orbit—and he returned from that second marathon in even better health than from the first.

But nobody has yet claimed it was easy. Maintaining muscle tone took the cosmonauts an inordinate amount of time, which detracted from productive work. The exercise was physically unpleasant since cleanup afterwards was difficult. Besides, many muscular systems did not seem to receive adequate workouts even with the hours spent every day.

Astronaut William Thornton, a doctor who worked closely with the Skylab experiments and who performed space sickness experiments in orbit in mid-1983, has criticized the effectiveness of current techniques for muscle exercise. "I can see little use for the Penguin Suit," he has said, "at least in its present configuration." Further, the Soviet treadmill onboard Salyut space stations is not fully effective because it only loads about two-thirds of normal Earth weight due to problems with the crewmembers' skin rubbing raw from the elastic harness. And lastly, claimed Thornton, the bicycle system is just plain useless: "Bicycles produce low repetitive forces which are totally inadequate to maintain the muscles for one-G walking," although it does provide enough work to keep up heart and lung capability. Concludes the astronaut, "One cannot take halfway measures and hope for the best— the bicycle can never maintain strength, vascular integrity, and neuromuscular coordination in the legs."

As an alternative, Thornton has designed and built a full-weight treadmill which has been used regularly aboard space shuttle missions. According to Thornton, such a treadmill could have a number of other important benefits—and here he runs into serious dispute with some other specialists.

One controversy deals with a well-documented postflight blood circulation weakness. Thornton suggests that it is not due to a heart weakness at all, but to the atrophy of leg muscles which through normal

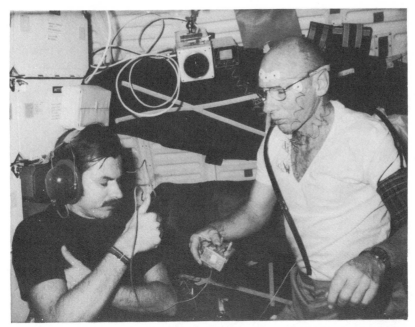

Dr. William Thornton *(right)* is America's leading expert on space sickness. He is shown here experimenting on a crewmate during a mid-1983 flight. (Photograph courtesy of NASA.)

walking activity on Earth help pump blood back up to the heart. "I feel that the musculature of the legs is crucial to venous return," he has written, and further that it is a decrease in this function which is "a primary cause of cardiovascular changes postflight. . . . If the leg muscles have suffered disuse atrophy, it is reasonable to expect that pumping capacity may suffer." Use of the one-G treadmill could block this leg deconditioning and solve the postflight blood circulation problem, asserts Thornton.

Bone Decay

Where Thornton really stands almost alone is in his argument that his treadmill (which does in fact have numerous improvements over prototypes tested on Skylab and over the best Soviet designs installed on their *Salyut 6* and *Salyut 7* space laboratories) can also combat what is probably the most insidious and worrisome effect of long-term spaceflight: the gradual dissolving of skeletal structure, called bone decalcification (or demineralization).

Careful measurements of the astronauts' diets on Skylab, together with X rays of bones preflight and postflight, showed that portions of the bone (particularly weight-bearing bones of the back and legs) were gradually dissolving in flight, much as do the bones of chronically bedridden invalids on Earth. The rate is difficult to measure precisely but it seemed to be as high as 2 percent of bone minerals per month. Since a 20 percent loss could be considered dangerous, such a rate could lead to bone failure on return to Earth after long orbital expeditions.

For a brief while, the Russians thought they had licked the problem. Early in 1980, Dr. Anatoliy Yegorov, deputy flight director for medical support at the Moscow Mission Control Center, issued some optimistic results from recent Soviet long spaceflights. "The amount of calcium lost during our 175-day flight was found to be approximately the same as the amount lost during our 96-day flight," he told newsmen. Chief Soviet space physician Dr. Oleg Gazenko reiterated those conclusions in an interview in *Soviet Life*: bone decay, he announced, was found to level off after a few months in space, and hence would definitely not be a problem for flights of nearly any reasonable duration.

Some American space doctors had received Soviet biomedical data as part of joint space medicine projects, and after examining the data closely they became very skeptical of the Soviet techniques used to measure these subtle physical alterations. The more data that was released throughout 1980, the more skeptical these Americans became.

Finally the two sides met at a scheduled colloquium in Moscow late that year. Dr. Joseph Sharp was part of the NASA team, and he recalled how the meeting went: "The Russians were very confident but not well supported," he recalled later. "We met in Moscow and quickly turned their whole glibness around. Their techniques for measuring [bone decay] were far from adequate." Sharp's negative conclusion was that "there is no useful decalcification data from any of the Salyut missions."

It turned out that the Soviets had been genuinely unaware of the insensitivity of their measurement techniques. Any bone decay which might have been occurring would not have been detectable (and the decay might have been leveling off, except that the Soviet data could not be used as an indication of such a trend). "When we pointed out the deficiencies," Sharp noted, "they accepted our criticisms and then invited us to participate in a monitoring program." Schedules for exchange visits were soon drawn up and accepted.

Possibly as a result, the Soviets put off all plans to extend their

185-day maximum duration mission flown in 1980. The following March, a few months after the NASA visitors had dropped their decalcification bombshell, Dr. Vladlen Vereshchetin of the Intercosmos Council officially announced that the USSR was temporarily calling a halt to marathon flights and that there would be no attempt in the near future to break the recently set space endurance record. The Soviet space spokesman told a news conference that the cosmonauts' work efficiency had declined after four or five months in orbit anyhow. Other Soviet space sources, including the record-holder cosmonaut Ryumin, spoke of a broad-based anxiety over medical and psychological problems, including bone decay and red blood cell shifts. In 1982, cosmonauts Berezovoy and Lebedev were launched on a planned 175-day flight to calibrate improved health maintenance procedures against the results in 1979. After five months in orbit, the mission was extended another month because things were going very smoothly, and some new medical ground was broken.

Bone decay, like space sickness, is caused by as yet undetermined factors associated with space conditions. Most researchers are confident that weightlessness has something to do with it. For Dr. William Thornton, the cause-and-effect is direct: "From clinical experience, bedrest studies, and flight experience, we know that removal of muscle forces and weight from bones causes a loss of minerals." The technical term for this is *disuse osteoporosis,* and in fact it affects a substantial number of people on Earth as well: paraplegics suffer from true disuse, while elderly people are subject to much the same decay for hormonal reasons. So the process has already been widely studied, mainly through bedrest experiments in which the bone decay progresses at much the same rate as in space.

Thornton's theory is that the stress forces associated with walking, on Earth or in space strapped to a treadmill, are the key to preventing disuse osteoporosis. Mere passive stress is not enough: "Only the large forces produced by activity of large muscle masses appear to prevent or reverse such losses," he argues. "Walking for one to one and a half hours per day under one-G forces might be adequate. At this time, it would appear that walking or jogging is the best candidate to prevent disuse osteoporosis."

Other specialists are highly skeptical. "It would take twelve hours of treadmill work every day," snorted one skeptic who himself favors the drug therapy approach.

And in fact the use of drugs has shown great success recently. At the Veterans Hospital in San Francisco, Dr. Victor Schneider and his associates, working on a NASA grant, maintained bedridden subjects (mostly skiing accident victims!) with a positive calcium balance over a period of several months. There was no sign of the normally expected bone decay. It was totally blocked due to a drug called dichloromethane diphosphonate (or dichloro MDP). These results were termed "very encouraging" by astronaut Dr. Anna Fisher.

Despite this, flight verification could be many years away since the Soviets have as yet shown no interest in the new drug, and they are the only ones planning manned missions lengthy enough to test the drug therapy under actual conditions.

"We have to keep in mind," noted space medicine researcher Dr. Daniel Woodard in Houston, "that bone demineralization is not a pathological process but rather an attempt of the body to adapt normal mechanisms to a changed environment. Because it is not really a disease, there may never be a cure in that sense of the word."

Under these circumstances, and at the current level of ignorance about space effects on the human organism, what might happen to true space colonists, people who intend to remain in space for the rest of their natural lives? How normal by earthside standards could those lives be, in a biological and medical sense?

Woodard has pointed out that bone decay is bound to ease off because it does on Earth in the case of paraplegics. "They are the only human model for the crew of an interplanetary spacecraft which might remain in space for years," Woodard pointed out. "The loss of bone does indeed stabilize, but only after the passage of years and after a degree of atrophy in the weight-bearing bones that would be irreversible and at least partly disabling"—disabling, that is, for life under full-weight conditions. Aboard a zero-G space station, the atrophied bones would be entirely adequate.

OTHER SPACE SIDE EFFECTS

Other concerns of very long-term human habitation of space have barely been examined. The cumulative effects of space radiations can be estimated, and adequate shielding should be able to accommodate this problem. Less well understood are problems associated with biochemical processes which take place in the absence of a strong magnetic field (such as on the Moon or in deep space); some disturbing data hints at potentially grave debilitating effects of such a condition, alien to life forms which have spent most of their evolution in the presence of such fields.

Questions of reproduction under space conditions are still topics for unbounded speculation. The only experiment in "space sex" involved rats aboard a Soviet bio-satellite on a three-week orbital honeymoon. None of the females became pregnant, but neither did any of their sisters in the control unit back on Earth—so the practicality of space sex is yet to be demonstrated. Embryo development under weightlessness should not be a problem, but Soviet experiments with quail eggs produced inconclusive results.

Some idea of the kinds of totally unanticipated problems which might crop up can be derived from a rumor which floated out of the halls of the Soviet-American space biology colloquium. One anonymous American doctor returned to America with an additional piece of data not part of the official record. According to this source, the Soviet doctors had determined that after several months in orbit the cosmonauts' sperm count dropped to zero (the prudish Soviets refused to discuss experimental techniques).

Billions of years of evolution have molded terrestrial organisms to thrive in and even depend on the conditions found near Earth's surface. Perhaps the near-infinite adaptability of the human mind and body will be able to accommodate the breakthrough into such a new environment as space. Perhaps science will have to replicate the evolutionary process to tailor-make organisms fitted for space conditions. In either case, a giant step in the evolution of terrestrial life is now in motion.

THE MEDIA'S NEED TO KNOW

All this is scant comfort for astronauts who are still getting sick on the space shuttle. In 1983, NASA went to emergency overdrive on the space sickness issue—and made a lot of enemies.

A fight nearly broke out at the press conference for the STS-7 astronauts in June 1983. One journalist demanded to know who on the crew got space sick, and to what extent. NASA spokesman John Lawrence deferred to a new NASA policy not to divulge such information unless in-flight activities were affected. The newsman stood up in the auditorium and demanded that the "cover-up" be lifted. Lawrence, who controlled the microphone, stood fast. The newsman grew abusive; Lawrence smiled bravely and asked for the next question.

The new policy was to a large degree the result of earlier crews' being forced to discuss in excruciating and embarrassing detail the degrees to which each of the astronauts had suffered from space adaptation syndrome. Many had found these public interrogations such a distasteful experience that they began to threaten to withhold all such information from everybody, doctors included. The astronauts argued with considerable justification that such disclosures violated the traditional confidentiality of doctor-patient relationships, so long as the information was purely medical and had no effect on their abilities to carry out their assigned tasks in space. NASA evidently agreed and implemented the new policy early in 1983. Newsmen who found that gruesome tales of vomiting astronauts made "good copy" were suddenly forced to write about more serious aspects of each flight.

All the same, space sickness was taken very seriously by NASA doctors. On the first seven flights, sixteen different astronauts were exposed to space conditions. Of them, five suffered loss of appetite, four had a general malaise, five had headaches, four had "stomach awareness," three had nausea, and six vomited. Others complained of other aches and pains, including backaches caused by readjustments of spinal muscles. And on several occasions, the astronauts' illnesses forced rearrangements of work schedules.

Since shuttle flights generally last about a week, several days of illness can impact the amount of work possible by each astronaut. Something had to be done.

NASA TAKES MAJOR STEPS

In 1983, NASA took two major steps. First, it added two astronaut-doctors to the STS-7 and STS-8 flight crews, with the assignment of studying in detail the physiological processes associated with space sickness. Secondly, in anticipation of the receipt of new data from this and other activities, NASA set up a new bureaucracy to coordinate studies on the problem and named an ex-astronaut to head part of it.

In June 1983, Dr. Norman Thagard was aboard STS-7, spending most of his time conducting measurements on himself. And in August of that year, Dr. William Thornton, 54, generally regarded as the world's leading authority on space sickness, made his own flight aboard STS-8. Both men were added to the originally four-man crews late in the training program; both men were supposed to be free of flight-related duties so that they could concentrate on the puzzles of space medicine.

And concentrate they did. Thornton brought up five lockers of equipment with him. Once it was set up on the shuttle's middle deck it looked, in the words of flight engineer Dale Gardner, "like an explosion in a Heathkit factory—with all the wires, cables, meters, dials, instruments, that Bill had floating round."

Thornton was ecstatic about his results. "I would compare the quality of the records and such that we obtained to those of labs on

Earth," he asserted. "Every measurement that I went up to make was done, and with excellent results. I tried not to waste a minute."

According to Thornton, space was "truly a unique environment for studying the human organism.. . . It had all of the benefits and possibly a few I didn't even expect."

Although he characterized the data as "exciting," Thornton was slow to jump to any conclusions about what might be the cause—or at least an effective countermeasure—of weightlessness in orbit. He did disclose one discovery: "It is a very, very different phenomenon from what one encounters in motion sickness on Earth"—despite the fact that the two syndromes share many common symptoms. Space sickness is not motion sickness, and treating astronauts for the latter has proven totally ineffective in alleviating the former.

The Space Adaptation Project

At the Houston Space Center, scientists have been mostly concerned with clinical approaches to the problem (i.e., how can the symptoms best be controlled). Actual basic research into vestibular and other physiological reactions to weightlessness and disorientation has been conducted primarily at the NASA Ames Research Center.

At Houston, a project called the Space Adaptation Project has been set up. Under it, a newly organized Space Biomedical Research Institute will consist of several subgroups. One is the Division of Space Biomedicine, with acting director Jack Schmitt, former astronaut and former U.S. senator.

According to NASA spokesmen, the technical objectives of the project fall into four basic areas. They are (1) countermeasures for prevention, (2) countermeasures for treatment, (3) prediction and choice of countermeasures, and (4) basic physiological mechanisms.

A Key Discovery

The push to conquer space sickness continued through 1984. Aboard STS-8, Thornton discovered that the body's upper gastrointestinal track shuts down for several days. This explains both loss of appetite and proneness to vomiting. Since drugs such as Scopo-Dex actually heighten this effect, tests were scheduled for a new series of drugs to counteract the actual physiological process.

7

Salyut 7 Breakthroughs

For many voyages, the most difficult part is the beginning. Other expeditions encounter the greatest difficulties during the midpoint. But for the cosmonauts of *Salyut 7*, in 1982, the ending was always expected to be the hardest part—and as it turned out, it was many times rougher than even the pessimists might have feared.

For a total of 211 days (a full month longer than the previous record), Russian space pilots Anatoliy Berezovoy (bare-ezz-oh-VOY, meaning "birch") and Valentin Lebedev (LEB-eh-deff, meaning "swan"), lived and worked aboard the new *Salyut 7* orbital outpost, more than two hundred miles above Earth. They circled the planet more than three thousand times and traveled eighty million miles. Both the duration and distance were nearly sufficient for a trip to Mars.

A ROUGH LANDING

Despite the record-breaking nature of the flight, perhaps its most noticeable characteristic was its routineness, its normality. The crew conducted both experiments and operational activities for industrial and economic applications back on Earth. They hosted two visiting crews, which included a French guest cosmonaut and the world's second woman in space.

All of those accomplishments were behind them when, on December 11, 1982, the two cosmonauts headed back to Earth. They had carefully loaded their *Soyuz T* spacecraft with the results of their long mission: logbooks, medical samples, photographs, videotapes, ampules of semiconductor and pharmaceutical materials manufactured in onboard devices. Next, they cast off from the *Salyut 7*, leaving it for use by subsequent crews, and fired their maneuvering engine to push them back into Earth's atmosphere.

Almost home, they ran into trouble. The unpredictable central Asian winter turned what should have been a routine nighttime landing into a near disaster when a cold front developed suddenly and enveloped the landing site in a fierce snowstorm. There was no turning back, however, as the *Soyuz* capsule plummeted eastward through the skies of southern Russia.

As later recounted by one of the cosmonauts, the landing was "very impressive" and "quite severe." The high winds brought the parachuting capsule down at an angle, and after hitting the ground it rolled several times down a slope. It came to rest on its side with the cosmonauts dangling awkwardly from their couches. It was then that they would have heard the howling wind and felt the chill of the blizzard creeping into their spacecraft.

Recovery helicopters were already in the air, and the capsule's radio beacon guided them to the landing site. A flashing light on the *Soyuz* was barely visible to the helicopter pilots, but when they tried to put down, their rotors stirred up so much snow they lost sight of the ground. Several landing attempts were called off, but as the *Soyuz* light flashed more and more weakly, pilot Nikolay Karasev swooped in and descended with enough forward motion to keep clear of most of the blowing snow. The dry streambed, where his helicopter crash landed, crushed his left landing gear. As doctors scrambled out, a second helicopter managed to land, guided by flares set by the first crew. All subsequent aircraft, however—including those with special survival gear and tents—were waved off.

It was twenty minutes after touchdown before the first rescuers knocked at the hatch of the snow–encrusted *Soyuz*. As the cosmonauts were leaving their craft, a cross-country tractor from the rescue forces managed to drive up. Since evacuation from the landing site was out of the question, the tired cosmonauts spent the rest of the night in the cab of the tractor. By dawn, the tractor's fuel supply had run out and

In 1982, cosmonauts Berezovoy *(left)* and Lebedev *(right)* spent seven months in space. Maintaining paperwork is always a major effort on space stations. (Photograph reproduced from a Soviet book.)

its heaters stopped, but the weather cleared enough for additional rescue helicopters to airlift the spacemen to more comfortable quarters.

Soviet officials called the recovery "perhaps the most difficult of all." This was hardly a fitting ending for the most successful Soviet manned spaceflight ever.

SPACE VISITORS

The new *Salyut 7* space station had been launched on April 19, 1982, and the two "marathon crew" cosmonauts followed inside a smaller *Soyuz T* spacecraft on May 13. A robot supply ship, *Progress 13* (essentially a modified Soyuz vehicle capable of carrying 5,000 pounds of supplies), linked up to the station on May 25; before casting it loose on June 6, the cosmonauts spent twelve days unloading it and pumping rocket fuel and drinking water into their station's tanks.

On June 27, a three-man ship, *Soyuz T-6*, was launched for an eight-day resupply visit; one of the crewmen was the French guest cosmonaut, Jean-Loup Chretien. After another *Progress* unloading (July 12 through August 11), during which the cosmonauts made a two-hour space walk to retrieve samples and test new tools, another visiting flight blasted off from Earth—and this time the space guest was a woman test pilot named Svetlana Savitskaya. The visitors left their fresh *Soyuz T-7* attached to the *Salyut* station and returned to Earth on August 27 inside the main crew's original craft, *Soyuz T-5*. Two more unmanned supply ships were launched, on September 18 and October 31. On December 11 the record-breaking cosmonauts returned home.

Life aboard the *Salyut* had been spartan but survivable. The cosmonauts' living quarters consisted of a main hall about ten feet across and twenty-five feet long, with small airlock sections at either end, where visiting spacecraft could dock. The whole *Salyut* weighed only twenty tons (of which two tons was scientific equipment), while the visiting *Soyuz T* and *Progress* ships weighed seven tons each.

SPACE EXPERIMENTS

In summarizing the accomplishments of the seven-month mission, Soviet space official Valeriy Ryumin reported that the cosmonauts had conducted at least three hundred different experiments. Some had been done only a few times, and some had been repeated every few days. They included work with materials processing, with biology, with earth resources surveys, with astrophysics, and with engineering tests intended for applications on future missions. The most important ones, according to the cosmonauts themselves, were those aimed at broadening the practical applications of spaceflight and creating measurable economic benefits to the Soviet Union.

"This is not only a scientific research base, but also a production base placed on a space orbit," boasted cosmonaut Lebedev at the postflight press conference on January 6, 1983. He repeated his claim: "For the first time, our space station undertook direct research production tasks."

Early in the mission, special small furnaces were sent up to the station inside Progress robot freighters. One was called the Kristall, and it was used to conduct metallurgical experiments aimed at obtaining alloys of high quality and purity. Another small electrical furnace was called the Splav, and it too was tested for several months. Visiting cosmonauts brought samples back to earth for detailed analysis.

Later, a new installation was put into operation. Called the Korund, the apparatus weighed 300 pounds and was apparently sent up in another Progress freighter. Whereas the first two furnaces had fixed programs for heating the samples, the Korund was controlled by a

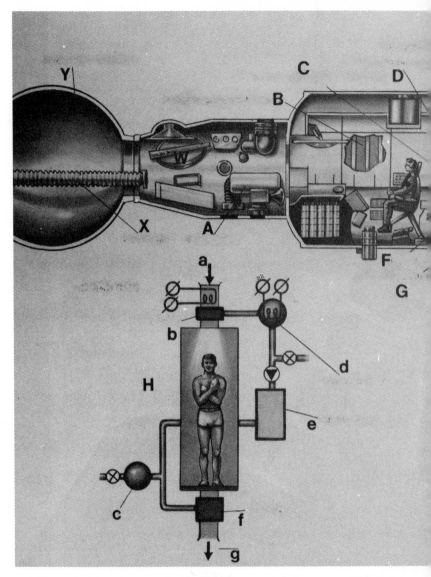

Cross section of *Salyut 6* systems. *(A)* spacesuits; *(B)* air regeneration; *(C)* water-recycling controls; *(D)* bicycle exerciser; *(E)* water-recycling apparatus; *(F)* water bottle; *(G)* CHIBIS vacuum apparatus; *(H)* shower; *(I)* shower controls; *(J)* treadmill; *(K)* mass-meter; *(L)* bunks; *(M)* trash airlock; *(N)* air filters; *(O)* scientific apparatus; *(P)* water bottles; *(Q)* toilet apparatus; *(R)* hygiene apparatus; *(S)* filters, towels; *(T)* air purifier block; *(U)* first aid kit; *(V)* personal grooming kit; *(W)* airlock hatch; *(X)* ventilation hoses; *(Y)* Soyuz at

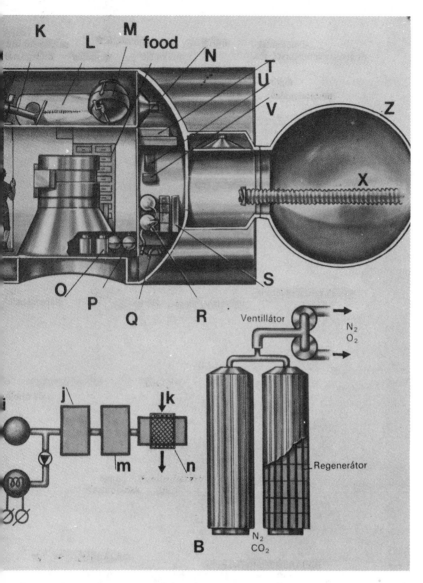

forward docking port; (Z) Soyuz/Progress at aft docking port. Shower details: (a) air intake; b) vaporizer; (c) waste water storage; (d) hot water tank; (e) water purifier; (f) condensator; (g) air exhaust. Water-recycling details: (h) hot water tap; (i) cold water tap; (j) water filter; (k) air intake; (m) moisture collector; (n) air filter. (Illustration reproduced from a Soviet book.)

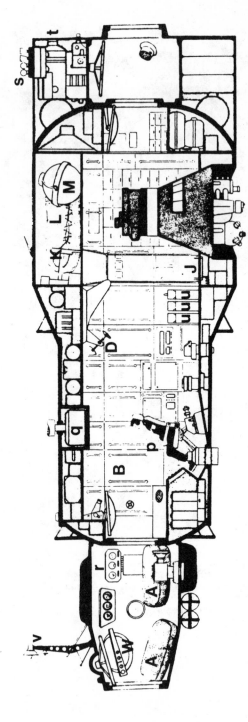

Salyut 7 appears to be extremely similar to its predecessor. (*p*) control station; (*q*) solar panel swivel mechanism; (*r*) airlock controls for EVA; (*s*) attitude control thrusters (four sets); (*t*) orbital adjustment engine (two); (*u*) multi-spectral earth survey camera; (*v*) rendezvous antenna (on boom). (Illustration reproduced from a Soviet magazine.)

minicomputer which took into account the actual conditions during the entire process. In the Kristall unit, the ampules (up to 10 millimeters in diameter) moved through a constant temperature field; in the Splav, slightly larger ampules were immobile while the temperatures changed. Korund, however, had three independent heat zones in which both the ampule position and the temperature could be varied at different rates. Samples weighing several kilograms, in ampules 30 millimeters in diameter and up to 30 centimeters long, could be processed; the unit had a carousel for a dozen such ampules. The temperature could range up to 1,270 degrees Centigrade, with an accuracy of half a degree; its automated program could run for almost three days unattended.

In the last weeks of their mission, the cosmonauts processed crystals of cadmium selenide, indium antimonide, and germanium sulfide doped with gallium. All are candidates for larger production facilities on future stations.

The scientist in charge of these experiments, Vladimir Khryapov, described their importance this way: "It is important that the possibility of obtaining big crystals of homogeneous structure has been proved," he told a newsman during the flight.

When such materials are produced on the Earth, the proportion of waste is rather high because all the efforts to obtain homogeneous crystals prove to be unsuccessful and the slightest violation of the homogeneity results in an electrical breakdown of the semiconductor element in one or another circuit, and this puts sophisticated and costly instruments out of operation.

"It is necessary to stress another aspect of the problem," Khryapov continued.

Research which is conducted in space makes it possible to understand better the mechanism of formation of heterogeneities in crystals and thereby improve technology in terrestrial conditions. The problem is that terrestrial gravitation and convection frequently obscure the course of processes, whereas in weightlessness they are seen more clearly.

Khryapov foresaw space factories in the near future.

First of all this will concern semiconductor material which does not have enormous weight requirements—on the order of hundreds of kilograms [per year]—but which is very expensive to produce on Earth. These may be, for instance, the crystals which are used in television technology, high-speed computers, medical equipment, high-capacity power transformers and other analogous devices.

Some unmentioned "analogous devices" include missile guidance systems and infrared tracking detectors such as those used on antiaircraft weapons—applications which give the Korund furnace a distinct strategic value.

Electrophoretic Separation Techniques

Other types of materials were also produced aboard *Salyut 7.* In August 1982, a visiting crew set up and started experiments with a unit called Tavriya. Its purpose is to manufacture high-purity drugs in orbit, through the use of electrophoretic separation techniques which on Earth are hindered by weight-induced convection mixing.

No such problems interfered with space operations, the Soviets boasted. After several months of operation, it was reported that purities were at least ten times better than those obtainable on Earth, and processing rates were several hundred times faster.

The material tested included human albumin, which was broken down into five components by two different methods, and hemoagglutinin, a particularly difficult protein to separate on Earth. Applications for other biologically active materials were also forecast, with the next step being "industrial biotechnology in orbit," according to one experimenter. Such substances can find widespread uses in pharmacology, genetic engineering, agriculture, and other industries, claimed another scientist.

Space Greenhouses

Besides growing crystals and pure proteins, the cosmonauts did some old-fashioned garden growing of their own. They planted and harvested at least a dozen different plants, and in the process succeeded with what had been frustrating Soviet space biologists for at least five years: they produced seeds in space.

Several different types of "space hothouses" were installed aboard *Salyut 7.*

In the Oazis 1A, the successor to units which have flown on spacecraft over the past decade, oats and peas were planted early in the flight and sprouted within a few weeks. Later, wheat and onions were also cultivated in the aptly named *space oasis.* By September, experimental subjects included garden peppergrass, dill, and kale; two months later the cosmonauts were growing more onions, plus parsley, radishes, and borage.

A unit called the Biogravistat contained a small centrifuge which produced artificial gravity to guide the growth of plants. By contrast, the Magnetogravistat provided an artificial magnetic field; flax sprouts produced small roots which grew along the directions of the magnetic field.

The most important results were obtained in the Fiton (Phyton) apparatus, in which seeds of the species arabidopsis (an herb from the cruciferae family) were planted. The unit provided a special nutrient medium, and it was isolated from the cabin air by special filters. More light was provided compared to the unit flown on *Salyut 6* several years earlier since those attempts to grow arabidopsis through its entire life cycle (seed to seed) had been unsuccessful; most plants had died, and those pods that had been produced contained no seeds.

But by mid-August the cosmonauts of *Salyut 7* reported the successful growth of full seed pods in the Fiton hothouse. The samples were brought back to Earth by a visiting crew, and biologists found about two hundred seeds in the pods. Eighteen were planted, and seven of them produced fully developed plants. Other seeds were saved for replanting on a future mission.

"Obtaining these seeds marks the solution of one of the key problems of space biology," claimed biologist Aleksandr Mashinskiy. "It remains now to perfect techniques for the cultivation of plants in space, to design new equipment, and to select other varieties of plants."

The future holds the promise of large space greenhouses attached to permanent space stations. According to explicit Soviet forecasts, plants grown in space will provide a significant fraction of future cosmonauts' nutritional requirements; other biological systems will help provide a "closed loop" life-support system where exhaled carbon dioxide is processed into rebreathable oxygen. The success of the arabidopsis experiments on *Salyut 7* may someday be considered one of the most important breakthroughs in space technology in this century.

Earth Resources Surveys

From its vantage point more than two hundred miles out in space, the *Salyut 7* provided a platform for observations of vast areas of the Earth's land and sea. The cosmonauts spent a substantial part of their working hours photographing the surface with a battery of special cameras as well as observing it directly with their naked eyes.

The primary instrument for these "earth resources surveys" was th MKF-6M camera array, manufactured at the Karl Zeiss Jena factory in East Germany. High resolution photographs were taken simultaneously in six spectral bands; later computer analysis on the ground was able to reveal details about agricultural use and health, about mineralogy and ground water resources, and about other information of vital interest to the USSR's marginal agricultural system. At least two all-day photo sessions were scheduled every week since more than six hundred Soviet organizations used the resulting space views.

Also aboard was the KATE-140 topographic mapping camera. It was designed to facilitate the surveying of poorly mapped regions of Soviet Central Asia and Siberia.

Eyeball observations were also important. From the Leningrad Institute of Metrology, the cosmonauts brought along a color atlas which contained more than a thousand different color samples. The human eye can differentiate up to 90,000 colors, and the cosmonauts were soon making recommendations on new colors they had seen; new pages were shipped up on supply flights throughout the mission.

The cosmonauts noticed a phenomenon described by earlier space-farers; their visual discrimination improved markedly over a period of the first several weeks in orbit. "Now we already notice much more than during the first days," Berezovoy remarked after a month in orbit. "In general we are starting to see that for which we were prepared."

One of Berezovoy's favorite observational assignments was that from the Soviet Ministry of Fishing. He and Lebedev were supposed to find biologically productive regions of the oceans, based on subtle color distinctions, a technique which had been proved out on *Salyut 6* missions in 1978–80. Over the months of the *Salyut 7* mission, the cosmonauts regularly and efficiently spotted and reported such regions all over the world's oceans. By the time they had landed, they had saved the Soviet fishing fleet more than twenty million rubles, according to a report from a Moscow organization for coordinating the exploration of natural resources from space. Catches were increased while sea journeys were reduced (saving fuel). An unnamed spokesman for the center issued a statement the following January in which he stated that "exploration of the earth's natural resources is one of the major tasks of the teams flying in orbit. In this way the expenditure on financing all the piloted flights in space has been fully recovered."

A catalog of the geological discoveries made by the first *Salyut 7* crew suggests that this last claim is no idle boast. The cosmonauts worked on twenty major geological projects during their flight, in particular the observation of ground structures difficult or impossible to detect from the surface or even from an aircraft.

According to Vladimir Kozlov, director of the USSR Ministry of Geology's Aerogeology Association, one of the initial discoveries by the cosmonauts was a hitherto-unexpected gas-condensate field in the lower Volga river basin. Also, a previously unknown chain of ring structures was discovered in the steppes between the Caspian and Aral seas, suggesting the possibility of gas and oil there as well. A structure called the Astrakhan Dome had its boundaries on the left bank of the Volga more precisely determined, providing information on another promising oil and gas field. Other discoveries included an ancient volcanic belt lying between the Chukchi Sea and the Sea of Okhotsk in eastern Siberia, some relatively small kimberlite pipes in eastern Siberia, unsuspected linear structures extending from the Caspian Sea to Lake Balkhash, and a new zone in eastern Yakutia believed to be rich in lead, tin, and other minerals. By the time the crew held its postflight news conference, six special geological expeditions had already been dispatched to the indicated regions in the Caspian-Aral-Balkhash area.

Kozlov had also claimed that such studies helped immeasurably in drawing up long-term state plans for economic development. Studies from space, combined with logical prospecting and surveying on Earth, cut the time needed to start tapping a mineral deposit by many times. The economic gain, Kozlov estimated, could be counted in the billions of rubles.

Astronomical Observations

Amidst all the "practical" exploration, the cosmonauts found time for some scientific research as well. This involved primarily astronomical observations.

The largest instrument aboard the *Salyut 7* was an X-ray spectrometer called the SKR-02M. The actual apparatus weighed almost half a ton and was located in the prominent "chimney" structure in the center of the station's widest section. With a recording area (its "beryllium eye") of 3,000 square centimeters and a 3-degree resolution, the device was repeatedly aimed at such targets as Cygnus X-1, at an interesting variable X-ray source in Ophiuchus, and at regions near the galactic center. Over twenty-five sessions lasting a total of forty hours had been made in the first five months of the flight. Of great interest, noted astrophysicist Yevgeniy Sheffer, was the observation of an unexpected burst that rose to a level thirty times higher than the normal value.

A small gamma-ray telescope, called Yelena, was turned on halfway through the mission. It could observe the flows of both gamma rays and charged particles (such as high-energy electrons in the van Allan belts). As described by astronomer Aleksandr Galper of the Moscow Engineering Physics Institute, the unit could be switched back and forth in its observing program by replacing certain units.

Highly sensitive astronomical cameras were also on board. The Piramig, provided by French scientists, was used to record phenomena in the constellations Andromeda, Cassiopeia, Cetus, Pegasus, Sagittarius, Orion, Taurus, and Auriga. On July 20, 1982, the cosmonauts were directed to photograph a comet; later, they used the camera to take images of emission layers in Earth's upper atmosphere.

RETURN OF THE COSMONAUTS

Of course, all of these activities may have kept the crew extremely busy, but they could not mask the fact that the most dramatic experiment on the seven-month mission was the survival of the cosmonauts themselves. Years before, Soviet space engineers had mastered the intricate technology of providing adequate air, water, food, air pressure, and temperature ranges to preserve human life in space. The challenges facing Berezovoy and Lebedev were twofold: the physiological effects of long-term weightlessness and the psychological effects of long-term isolation.

Throughout the mission, there were occasional hints of bouts of interpersonal tension aboard the *Salyut 7*. These were usually blamed on accumulated crew fatigue and occasional frustrations over equipment problems. After landing, Berezovoy was asked whether or not he and Lebedev ever grew tired of each other. He candidly answered that they had had to "overcome psychological difficulties." Further, the cosmonaut urged that participants in future prolonged missions should be more thoroughly prepared psychologically for the experience. In the final months of the flight, the crew's efficiency declined markedly, probably for psychological reasons.

Physically, the cosmonauts returned from space fatigued and weakened but essentially healthy. Through a vigorous exercise regimen they had fought off the seductive adaptation to weightlessness which, through the weakening of muscles and the heart in particular, might otherwise have threatened to make the stressful return to terrestrial conditions a serious risk to their lives. But tests back on Earth showed no irreversible body changes. Flights of even greater duration were confidently predicted.

8

Soyuz T-8 Rendezvous

With more than the usual tension, the two cosmonauts guided their *Soyuz T-9* spaceship through the intricate series of maneuvers required for an orbital rendezvous. Ahead lay their target, the *Salyut 7* space station and the add-on module called *Kosmos 1443*. It was the biggest space complex ever approached by cosmonauts in orbit. If they made the linkup, it would certainly be a day to remember.

What lay behind was equally memorable: their predecessors in orbit, only nine weeks earlier, had tried and failed to make a rendezvous with the same target. That setback had been embarrassing, but another failure could be catastrophic.

But nothing went wrong. At 2:46 P.M. Moscow time on June 28, 1983, cosmonauts Vladimir Lyakhov and Aleksandr Aleksandrov linked their *Soyuz* craft to the aft end of the *Salyut*. The rendezvous and docking had succeeded!

A SPACE JINX?

For the superstitious, the failure of Russia's *Soyuz T-8* mission the previous April 21 proved out the power of an old jinx. Even though the embarrassing foul-up frustrated briefly the Soviet drive to set up a permanent space station, foreign observers were confident that a new attempt would be made—and would almost certainly succeed—within a few months. And so it was, when *Soyuz T-9* was launched June 27. There have been such failures before and they have always been only temporary setbacks. If the jinx is authentic, there will be more such occasional failures in the future.

In almost every case in the previous sixteen years, the first rendezvous mission assigned to a member of a new cosmonaut class had failed. Every rendezvous mission failure, of which there were ten, was marked by the presence in the crew of a representative of a new group of Soviet spacemen. Of course it is a coincidence since pilot error has never been a significant factor (the Soviets allow their space pilots very little latitude to exercise either skill or clumsiness in flight). But all the same the pattern is odd (and it can be applied to American astronaut classes as well!).

This time it was the "fault" of mission commander Vladimir Titov. The 36-year-old rookie spaceman had been selected as a cosmonaut in 1976, and he was the first man of that group to be assigned a space mission. He was accompanied by two veteran civilian cosmonauts, Gennadiy Strekalov and Aleksandr Serebrov.

The mission had been anticipated for several weeks, ever since the *Salyut 7* space station had linked up with the *Kosmos 1443* module. This doubled the habitable size of the spacecraft and provided a significant improvement in instrumentation and electrical power supply. Early in April 1983, the forty-ton complex changed its altitude slightly, a long-recognized characteristic of linkup preparations. One launch window on April 11 was passed up (reportedly, a countdown had been under way, but was cancelled), and then a second on April 14. New orbital maneuvers followed, and the window for April 20 was calculated precisely to the minute by Western spacewatchers.

Right on time, the *Soyuz T-8* blasted off. It was 5:11 P.M. Moscow summer time. An hour later, a Soviet official made the announcement.

The Soviet news media at first played the story down, evidently saving superlatives for after the docking when the new modular station would be manned for the first time. *Izvestia* even carried the launch headline and crew portraits "below the fold," demonstrating a nearly unprecedented low-key attitude. The cosmonauts, who were using the call sign "Okeyan" ("Ocean") in flight, had made some general comments on their planned experiments aboard the *Salyut 7/Kosmos 1443* complex. One particular boast by Gennadiy Strekalov would soon seem a cruel irony: "We have long been getting ready for the flight," he told

In April 1983 the *Soyuz T-8* crew (*left*, Aleksandr Serebrov, researcher; *center*, Vladimir Titov, commander; and *right*, Gennadiy Strekalov, flight engineer) failed to rendezvous with the *Salyut 7* space station. Serebrov was dropped from the crew to allow their next attempt to carry more propellant, but Titov's and Strekalov's space vehicle (it should have been called *Soyuz T-10*) blew up on the launchpad in September. When a crewman from the *Soyuz T-11* was medically disqualified six weeks before blast-off, Strekalov was transferred to that mission.

a TASS news agency correspondent, "so now all three of us can say with confidence that we are ready for the encounter with the station and we have not the slightest doubts as to the reliability of our space equipment." If Strekalov spoke the truth, it was a good excuse not to study backup flight procedures, an omission which he and his shipmates were shortly to bitterly regret.

Using the same prediction techniques which had pinpointed the moment of launch, Western experts calculated that the linkup would occur at 6:30 P.M. the following day (mid-morning in Washington, DC). Allowing an hour for the announcement message to make its way through the Moscow censorship bureaucracy, these experts expected to hear of the successful linkup by midday Washington time.

But no announcement came at that time, or for hours afterwards. In fact, it eventually took Moscow twelve full hours to confess to the world what the foreign observers had already begun to suspect: the mission was a failure and the crew would have to return to Earth. But

Soviet artist's concept of 1983 rendezvous of *Soyuz T-9* with the *Salyut 8* and *Kosmos 1443* combination. By grisly irony, the background of this cosmic scene is southern Sakhalin Island, exactly the spot where the Korean airliner was shot down later that very same summer.

even that brief announcement did not reveal the extent of the trouble or the danger facing the men in orbit.

LINKUP FAILURE

Even before the official Moscow admission, Western news stories had described the chain of events. The *Soyuz T-8* had approached quite close to the *Salyut* target satellite and had remained fairly close for many hours afterwards. But it seemed that the two space vehicles never completely matched their velocities through space, thus making a safe physical connection an impossibility.

Speculation was that the fault lay with the new guidance system installed in the totally redesigned *Soyuz T* series. When the new vehicles began manned space missions in 1980, Soviet space engineers boasted that the onboard computer could plot a course through space and achieve an orbital rendezvous with far less rocket fuel than needed by the old-model Soyuz. Additionally, the *Soyuz T* contained what was called a "unified propulsion system," in which fuel for the small attitude control rockets and for the large orbital maneuvering rocket came from the same set of tanks. This was listed as a great advantage because several earlier rendezvous missions had been called off after the attitude control fuel ran out, even though large reserves remained in the maneuvering engine's tanks.

Guidance during the final approach to a space target is a very delicate and complex process because "everyday experience" on Earth or in aircraft will persuade a pilot to make the wrong maneuvers in space. That is why computers, fed with precise measurements of range and relative velocity, must be used for efficient computation of correct course changes.

But the computer on the *Soyuz T* had earlier performed poorly. On *Soyuz T-2,* in June 1980, the automatic guidance system failed when the ship was only 180 meters from its target, making the final manual approach "relatively complicated" (in the modest words of one of the cosmonauts). On *Soyuz T-6,* in June 1982, the computer began generating erroneous directional commands at a range of 900 meters, forcing the cosmonauts (including Jean-Loup Chretien) to take manual control and complete the docking.

"The computer on board stopped doing automatic piloting," recalled Chretien, who went on to describe how the *Soyuz*'s guidance computer "left the field" (a pilot's term suggesting a state just short of total failure). "The vessel was no longer oriented towards the station," he continued. "We could no longer see the station. There was rotation on three axes due to instructions to the onboard computer to go back to the original position. Since it was at that moment the computer broke down, the vessel was in rotation around its three axes and was thus like a stone rolling over." The crew took over manual control and were able to complete the docking maneuvers.

That had been the previous June. This time, suspicions were aroused even before the scheduled docking. The Associated Press bureau in London began circulating a story early on April 21 (even before the cosmonauts had woken up for their first workday in orbit) that there were control problems. Later, a Soviet newspaper admitted that "deviations" occurred early that day and that ground controllers had to "change the plan of maneuver." Another Moscow paper referred to "unpleasant surprises" which greeted the cosmonauts.

Months later, spacecraft commander Titov (no relation to the 46-year-old retired spaceman Gherman Titov, in 1961 the second man in orbit, who has since assumed the traditional appearance of an over-

weight Russian general officer) wrote a remarkably candid article about
the failure. It appeared in *Red Star* newspaper on August 9.

Barely two hours into the flight, disaster had struck. The boom
with the radar dish had not fully deployed, precluding its operation.
The cosmonauts rocked their spacecraft back and forth in a futile effort
to shake loose the jammed boom. "To be honest," wrote Titov, "we
were not happy."

The men stayed up late trying to figure out a fix, but were finally
ordered to sleep by Mission Control. "I slept for about three hours,
in fits and starts," Titov wrote. "The complicated situation went round
and round in my head." In a major confession of preflight training
inadequacy, he admitted, "What we had encountered in the actual
flight was nothing like any of the unplanned situations known to us. . . .
We were now moving onto a completely untrodden path."

Although the linkup should have occurred after 25½ hours of flight,
the *Soyuz* was still approaching the *Salyut* many hours later. At 26½
hours, the ships were about 4,200 feet apart; four hours later, they
were less than 1,700 feet apart but swinging back and forth in a complex
uncontrolled maneuver. A few hours later, according to a schedule
released the next day in Moscow, the cosmonauts were ordered to call
off the rendezvous and prepare for return to Earth.

Later, a short but exceptionally candid article in the daily newspaper *Sovetskaya Rossiya* gave an explanation of sorts for the cancellation: "What specialists call an irregular situation arose," correspondent Boris Gerasimov wrote. "Owing to malfunctions of a system
[the radar, as it turned out], the spaceship could not go into the pre-
scribed approach pattern. The most diverse variants were examined
and complicated attempts to enter the proper zone were undertaken
both on the ground and by the crew. Nevertheless, the attempts to
perform this operation were not successful." Following several hours
of frustration, "the flight directors, after analyzing the situation, cancelled the docking and proposed preparations for terminating the space
mission." Amidst all of these words, however, no details were given
at that time about exactly what system had failed.

The *Soyuz* was also observed by Western tracking agencies to go
frequently into modes where it was slowly rotating, once every two
minutes. Although this mode had been used in the past, it was only
for periods when the ship's attitude control system was turned off—
or not operating properly.

The exact maneuvers to be followed in the rendezvous can certainly appear complex. When the *Soyuz* was launched, the *Salyut* was
in its 5,785th orbit of the Earth, ranging between 181 and 189 statute

miles with a period of 90.34 minutes. It crossed the equator at 17.4 degrees east longitude on a course that took it past the launchpad about two hundred miles to the east. Twenty minutes later, the *Soyuz* blasted off, after waiting for Earth's rotation to carry the launchpad into the orbital plane of the space station.

Soyuz T-8's initial orbit was lower than *Salyut*'s, and thus had a shorter period of 88.5 minutes. Since the *Salyut* was twenty minutes (or, at an orbital velocity of five miles per second, six thousand miles) ahead, the *Soyuz* was catching up at a relative velocity of about 400 miles per hour. But this approach speed, which could overtake the *Salyut* within fifteen hours, is not maintained for long. On the fourth pass around the earth, the cosmonauts fired their rocket engine to raise their apogee (or highest point) nearly to the altitude of the *Salyut*. This increased the orbital period to 89.5 minutes and the approach rate to less than 200 miles per hour. The following day, on the seventeenth orbit, the *Soyuz* began a final sequence of rocket burns to raise its orbit to the *Salyut* altitude at precisely the moment when the range was smallest. Most of those burns went correctly; by the time trouble came up, the two space vehicles were in nearly matching orbits with a relative velocity of only several miles per hour. While close, that wasn't close enough: navigation problems got no easier as the range dropped; instead, they became more complex and unforgiving of error. And there was such an error.

Titov described exactly what had gone wrong, in his "true confessions" published months later. By ground guidance, the *Soyuz T-8* was brought to within a mile of the *Salyut* complex: "After that we would have to maneuver ourselves." But without radar, the crew would have to use the target's angular size as a means of estimating range.

The three cosmonauts read off the size in their viewing screen (a periscope which projected the view onto a screen incised with angle hacks), and Mission Control told them how long to burn their rocket engines. This worked to a range of 900 feet when the *Soyuz* passed out of radio communications range. They had hoped to be closer more quickly, but Titov flew very cautiously. Now he was totally on his own in a failure mode he had never practiced. To make things worse, the sun now set and only the *Salyut*'s running lights were visible.

"Approach velocity seems fast," he recalled thinking. "I switch on the braking engine." At a range of 500 feet, where rendezvousing American spaceships are creeping at 2 feet per second, he thought: "Velocity still seems quite high. At night this is dangerous. Perhaps we will collide. I burn the engine to take the vehicle downward. We are flying past the station." The rendezvous was a failure.

By sunrise the distance had risen to 2–3 miles. "The error in range and velocity in the approach was too great and this made it impossible to dock." Mission Control ordered the men to prepare to land: "The amount of fuel remaining on the vehicle would not permit another attempt," Titov sadly reported. "This experience was a difficult one," he concluded.

In Kettering, England, veteran spacewatcher Geoffrey Perry was tuned in to the cosmonauts' voice communications with Moscow Mission Control. Six hours after the scheduled docking time, they were still firing their rocket engine, Perry reported. One conversation had the crew saying, "Received three burns, zero zero one six, zero zero one seven, T-R-P zero zero two zero, on leaving the shadow. I'll give the command Boris-Four. Command entered." Perry, who watched the two craft pass over England at dusk on April 21 (since the *Soyuz* was fifty miles behind the *Salyut,* it must have overshot the target and gone into a higher and slower orbit), observed that "they weren't their usual normal jolly selves—they were very terse."

RETURN TO EARTH

Once the mission cancellation was announced, Western observers were able to consult their orbital charts again and pinpoint the next opportunity for landing. That would occur at "about 5:30 P.M. Moscow time," one expert estimated. The actual touchdown occurred at 5:29 P.M. It was the world's shortest-ever nine-month spaceflight, one cynic joked.

But it had been no joking matter. As the ship was coming in for a landing, the top Soviet government leaders were attending a mandatory Lenin ceremony (it was the anniversary of his birth). At 5:31 P.M., an aide rushed to the podium, in full view of the live television cameras, and gave a note to Defense Minister Ustinov (who has long been the overall director of Soviet space activities). He read it, smiled broadly, and slapped the table—and then passed the note to Andropov and Gromyko, who also broke into grins of relief.

Evidently, there had been serious doubts that the cosmonauts would survive their return to Earth. A plausible explanation is that the frantic and drawn-out maneuvers the previous day had exhausted not only the attitude control fuel allotment but a significant fraction of the maneuvering fuel allotment as well, which would have threatened a loss of control during the searing plunge back through the upper atmosphere. This was possible because the reentry tanks could be tied into the maneuvering tanks, a feature once touted as a "major advance"—

but this time, the crew may have carelessly used too much of the propellant normally allocated for the return journey. If that had happened, the spacecraft and the three men would have been incinerated— that anxiety may have been weighing on the minds of Andropov and his colleagues, and upon the desperate engineers at Moscow Mission Control. But fortunately that tragedy did not materialize.

The embarrassment of *Soyuz T-8* is therefore, in perspective, much less than it might have been. And within two months the new mission of *Soyuz T-9* erased the bitter memory of the temporary setback (the Salyut fired its own onboard rockets a week after the rendezvous failure, raising its altitude to await the new launch attempt). If the mission can serve any beneficial purpose it is to remind the world that spaceflight is not yet entirely routine, and that courage and ingenuity remain the absolute requirements for those who would venture into such orbits.

And besides, as it turned out, the rendezvous failure was far from the worst manned space setback of 1983.

9

Modules in Orbit

On Earth there are a multitude of reasons why a homeowner might decide to add on an extra room. The family might have grown too large, or new activities might have begun which required their own roof, or additional storage area might be needed.

The same is true in orbit. As the twenty-ton three-room Soviet Salyut space stations were being utilized to their fullest in the late 1970s, Russian spacecraft engineers realized that the Salyut module's basic limitations had to be overcome by hooking on an entirely new module. On the *Salyut 6,* both working space and electrical power were severely limited. Many pieces of often-used equipment had to be built in fold-out form, for laborious and time-consuming unstowing and re-stowing. The Salyut and the seven-ton Soyuz manned capsule (and the latter's one-way unmanned cargo variant, the Progress) remained limited in the amount of cargo which could be returned to Earth at the end of the cosmonauts' mission: a cannister holding less than one hundred pounds of logbooks, samples, exposed films, magnetic tapes, and other results was all that could be brought back from months-long expeditions.

THE EXTRA ROOM

So when cosmonauts Vladimir ("Volodya") Lyakhov and Aleksandr ("Sashenka") Aleksandrov finally arrived at *Salyut 7* in mid-1983, they found an "extra room" docked to their orbital home. The spacecraft, called *Kosmos 1443*, had been launched the previous March and had automatically linked up to the *Salyut*'s forward docking port a week later. The cosmonauts parked their *Soyuz T-9* at the *Salyut*'s unoccupied aft docking port and moved on board. Two days after arrival, it came time to enter the extra room.

"Expect some surprises," radioed Moscow Mission Control cryptically, as the two cosmonauts worked at the hatch mechanism.

Soon the hatch was open, but Lyakhov paused, perhaps pondering the implications of the good-natured warning. Mission Control demanded to know why the delay, and Aleksandrov tried to explain.

"He sweated it for about fifteen minutes," the civilian flight engineer reported.

"Who was sweating, was it Volodya?" asked Mission Control.

"No, not Volodya, the hatch," came the answer. "But Volodya probably was, too."

"Well, take your time about going in," was the new advice. "The air could be stale. Do you smell anything?"

Lyakhov sniffed the air at the mouth of the transfer tunnel. "Just a slight smell of paint," he reported. "The smell is normal, technical-smelling." Spotting a television camera which had been trained on the entrance to the module, he glided into the tunnel.

"It's very spacious in here!" he exclaimed as he turned on the lights. "It's just fine here! Sashenka, come on in." The two cosmonauts were delighted with the new room. "It's easy, beautiful—there is room for work in here. We can carry everything all about here."

On the wall they saw the "surprise," a large hand-lettered banner. "Welcome!" it greeted, "Pull towards you, not sideways—good luck with the unloading."

The message had been written by Vladimir Titov, who had commanded a mission in April which was aborted when the rendezvous radar dish antenna failed (see Chapter 9). He had expected to read the sign himself, with his own crew—but now he was at Mission Control, cheering on his replacements.

Lyakhov was pleasantly surprised by the new room's layout. "Here in space it has turned out to be much more comfortable and attractive than when I became acquainted with it on Earth," he confessed.

"You have almost three tons of dry freight," Mission Control

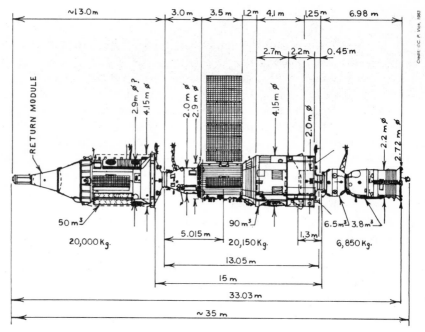

Line drawing of the Salyut and its module. Interior arrangement is still uncertain. (Illustration courtesy of C. P. Vick © 1983.)

reminded the excited cosmonauts. That calmed them down: "Yes," they replied, "there's a little bit of work here."

The cargo inside the module included many of the same kinds of items brought up in the past aboard Progress freighters—except that *Kosmos 1443* carried 2½ times as much. There were food, water, air regenerator cannisters, filters, replacement equipment, sports and training equipment, space instruments, medical equipment, movie and photographic materials . . . and a three-string guitar. Also, here and there were stashed surprise souvenirs for the crew, including some packets of letters from relatives of the men who were supposed to have boarded the module two months earlier. These letters, unopened, were subsequently returned to Earth.

The cargo was packed in cylinders for ease in transfer and stashing aboard the already-cramped *Salyut*. A small unfoldable track system ran through the hatch, and each cylinder could be trolleyed down the track from one room to the other.

Kosmos 1443 was a special kind of spacecraft, indeed. In form and size it approximated the *Salyut 7* itself, being 43 feet long, 13 feet across at its widest point, and having two solar panels with a wingspan of 53 feet and an area of 400 square feet (providing three kilowatts of power). The module weighed as much as *Salyut,* 44,000 pounds. Its internal pressurized volume was 50 cubic meters, half that of the *Salyut-Soyuz* complex.

Besides storage space, the module contained its own propulsion system and its own landing capsule. The former system, according to Lyakhov at the postflight press conference, "made it possible to maintain the whole complex in orbital alignment for a long time, which is necessary for the conduct of a number of experiments" (the module's engines also raised the altitude of the whole complex several times). The latter system relieved the bottleneck in space-to-Earth transportation which had in the past severely limited the amount of materials returned at the end of the mission.

By mid-July, two weeks after reaching the incremented space station, the two cosmonauts had made as many surface photographic runs as had the previous year's expedition in seven months. A profligate use of attitude maneuvers and alignments was possible using the half ton of excess propellants in *Kosmos 1443*'s tanks. Additionally, since the landing capsule could bring back up to one thousand pounds of material, there were no longer any restrictions on exposing as many rolls of film as possible.

Destruction of the *Kosmos 1443*

Although many Western observers expected the new module to provide a permanent expansion to the *Salyut,* after its own landing capsule had been jettisoned, presumably to expose a new docking port for subsequent visits, such was not to be. The cosmonauts loaded the landing capsule with 770 pounds of cargo, and on August 14, 1983, the entire *Kosmos 1443* separated from the *Salyut.* Nine days later the landing capsule was in turn jettisoned from the free-flying module, whence it landed successfully in Soviet Central Asia. The *Kosmos 1443,* meanwhile, maneuvered carefully toward what observers suspected would be a redocking with the *Salyut.* But again, surprise and disappointment were experienced when on September 19 the module was deliberately steered back down into the atmosphere, where it burned up.

Later, *Aviation Week* magazine reported that its sources within the American intelligence community had detected a "serious malfunction" aboard the module. Alternately, the destruction of the module followed the September 9 near-catastrophic fuel leak during the transfer of highly volatile propellants from the *Progress-17* tanker into the *Salyut 7*'s tanks. Reportedly, half the *Salyut*'s control jets and two of its three sets of propellant tanks had been lost (the Soviets at first denied this, but months later grudgingly admitted that a "small" leak had indeed occurred as reported). Since such a serious loss of *Salyut* propulsion and attitude control capability would have made any subsequent re-rendezvous impossible anyway, the decision may have been made to dump the module.

MULTIMODULAR STRUCTURES

Whatever the disappointments of *Kosmos 1443* in 1983, its successors will be playing a vital role in Soviet manned spaceflight. Academician V. Avduyevskiy, a leading Soviet space scientist, described some possible follow-on configurations in mid-1983. "Instead of a freight reentry apparatus, vehicles can carry modules—observatories, greenhouses, smelting shops producing metals or semiconductor crystals with properties unobtainable on Earth—and at the same time play the role of space tug, assembling complex space structures from individual elements."

Fanciful Soviet view of future modular assembly of large space station based on Salyut.

Early in 1984, chief cosmonaut Vladimir Shatalov issued a remarkably explicit prediction: "The modular execution makes it possible to extend the facilities of orbital scientific complexes," he told Soviet journalists. "A module can be special purpose: an astronomical observatory, a geophysics laboratory, a greenhouse for biological research and experiments, a technological workshop. . . . In short, there are many different versions."

Shatalov continued: "Eventually I can imagine the future of Soviet astronautics—this is space complexes composed of separate units. There will be everything on them for normal life, work, rest and sports. The units will contain research equipment, laboratories and workshops. I think that some of the units will remain aloft for some time in an autonomous flight without communication with the station, according to a set program. They will be docked then to the basic station for repair, exchange of equipment and supplies of medical preparations, extrapure materials, etc."

Two-time space veteran, civilian flight engineer Aleksandr Ivanchenkov added another view: recent experience "enables us to look ahead to the creation of multimodular structures. In the long term the appearance of the station will be a single unit with a large number of docking ports to which will be docked various modules performing different functions."

Top Soviet space designers have expressed their total satisfaction with the expanded capabilities of this new system. Nothing bigger is being planned because nothing bigger is needed in the near future, they say.

In mid-1983, ex-cosmonaut Konstantin Feoktistov, now a leading spacecraft design bureau official, was interviewed by a Russian journalist on the question of future space stations. "I expect it is worth considering the question of increasing their size, is it not?" he was asked innocently.

Feoktistov barked back, "Why?" The journalist was flustered. "It would offer new facilities," he suggested weakly.

"How? Name them!" Feoktistov demanded. "Prove their urgency, and prove they cannot be realized on stations of the present size—and I will be the first to request a budget to develop the new equipment."

If no further hardware development can be justified to Feoktistov, the extra rooms provided by the *Kosmos 1443* family must have plenty of applications. If so, it will take many years to work out all of these applications fully. There will be lots of extra rooms flying up into orbit!

10

Drama of Soyuz T-10A

The landing of the *Soyuz T-9* spacecraft on November 24, 1983, was almost disappointingly routine. The two cosmonauts, Vladimir Lyakhov and Aleksandr Aleksandrov, had spent only 150 days in space, shorter than three earlier Soviet orbital expeditions and hardly longer than a Salyut mission five years before. For the official Soviet news media, it had been a low-key "ordinary" space mission, but for the Western news media it had often been the topic of frantic speculation.

Hundreds of miles above the Earth, possibly the greatest spaceflight drama since the struggles to save *Apollo-13* and the crippled *Skylab* space station had been unfolding in the weeks leading up to the landing. Two other Soviet cosmonauts had narrowly escaped death when their launch rocket blew up underneath them. Lyakhov and Aleksandrov had remained in orbit long past their scheduled return date, and widespread reports had it that the men were marooned. Their orbital habitat, the space station module *Salyut-7,* had also been subject to dramatic rumors, most critically concerning a possible rocket fuel leak which may have crippled most of the station's maneuvering capability. Just before their landing, the cosmonauts performed two space walks to add extra solar panels to their station's original power wings, thus giving support to still other rumors about electrical power problems.

A day before the successful return to Earth, the flight director at Moscow's Mission Control Center had dismissed all of the fuss. Valeriy Ryumin, himself a three-time space veteran and the orbital endurance record holder, responded to the statement by Soviet journalists that "certain of the mass media in the West have decided to fantasize" about problems by asserting: "These are just false rumors." Regarding the reports of the potential unreliability of the overage *Soyuz T-9*, Ryumin claimed that "if we did have any doubt, then of course we would not be preparing for any landing—we would be preparing a rescue craft."

Ryumin was not being entirely candid although it was clear that Western anxieties over the condition of the *Soyuz T-9* were highly exaggerated. The landing time alone clearly indicated that his assertions about the utter normality of the mission were equally exaggerated; Soviet manned flights are launched at an appropriate date so as to be able to make their landing—after a nominal mission duration—at late afternoon local recovery zone time. Each day's best landing time is on average about twenty-four minutes earlier than the previous day's time, due to the precession of the station's orbital plane. The *Soyuz T-9*, having missed its normal landing date by more than 35 days, came down at 2:00 A.M. Something, at least, had not gone as planned.

And through it all, the Western news media struggled, with notable lack of success, to make sense out of these developments in the absence of any official Soviet statements—but in the dubious presence of "leaks" from "informed sources" in Washington and Moscow. Often the small pieces of the puzzle do not even fit each other. Yet many of the leaks were grudgingly confirmed by Soviet experts, in private and at public press conferences (unreported in Soviet newspapers), in the following months.

THE PLANNED CREW EXCHANGE

To make sense out of it all and place it in proper perspective, it is necessary to understand just what the Soviets were trying to do in space in 1983. Since only their failures or occasional propaganda "space spectaculars" tend to make the newspapers and the evening television news, this task can be difficult.

What the Soviets attempted to do on September 26, 1983, was nothing short of making space history with a new "giant leap for mankind." On the flight program of cosmonauts Vladimir Titov and

Gennadiy Strekalov was the first-ever exchange of space crews. They were to have been the world's first space station relief mission, allowing the two previous inhabitants of *Salyut-7* to return home while the space station continued to function.

Always in the past, long-duration crews have flown to an empty station, activated it, received visiting crews, then after many months deactivated the station and left it empty. That was not supposed to have happened this time, but it did. However, one thing is certain about Soviet space operations: they have come to accept a certain fraction of failures, have learned how to live with them, and will certainly try again.

These former visiting crews spent about a week aboard the *Salyut*, already occupied by the main crew. In September 1983, however, the overlap may have been scheduled to last two or three weeks. This would have given the new crew sufficient time to become acclimatized to space conditions and to be fully briefed by the departing crew on the routines and idiosyncracies of the onboard equipment. Some support for this hypothesis is provided by the fact that the aborted launch attempt (it was supposed to have been called *Soyuz T-10*) occurred at least ten days earlier than the standard orbital conditions for a routine week-long visit.

The purpose of such routine week-long visits is tied in with another problem to which the September launch failure led. The seven-ton Soyuz craft, unlike the multi-year-mission twenty-ton Salyut module, has a markedly limited lifetime in orbit. As a general rule, such spaceships only spend sixty to eighty days in orbit before returning. Three times before, Soyuz ships had remained in space slightly in excess of 100 days, but all three were contingency cases.

So during long space station missions of six or seven months duration, the smaller Soyuz craft attached to the larger Salyut must be periodically swapped for a fresh one. About every fifty-eight days (the period over which the station's orbit precesses through an entire twenty-four hours and hence presents the same orbital lighting conditions), a new visiting crew is launched. Since their mission is extremely simple, guest cosmonauts from other countries can be sent along for propaganda purposes, at no risk to mission efficiency. In 1982, this program included a French pilot and, on the next mission, a Russian woman pilot. In 1984, an Indian guest cosmonaut was flown.

The reason for this empirical duration limitation rule has never been directly addressed by Soviet space officials, but reasonable analogies with American space vehicle characteristics can provide some insight. First among them is the presence of highly corrosive hypergolic propellants in the Soyuz service module tanks; as weeks pass, the exposed valves and seals in the engine could be expected to deteriorate and lose their integrity—they may suddenly leak all the fuel, or they may not even function at all when called upon during the critical return to Earth. In addition, electrical batteries aboard the Soyuz may slowly lose their charges. Lastly, the simple exposure to sixteen cycles of sunrise/sunset thermal stress each and every day may have a cumulative impact on the mechanical hardware of the ship.

In December, Moscow Mission Control Center's chief flight director, former cosmonaut Valeriy Ryumin, confessed that at the time of launching in June the safe lifetime of the *Soyuz T-9* was considered to be only 120 days. Subsequent analysis and ground tests demonstrated that 150 days would be feasible. Following the landing, some jubilant Soviet space officials were even saying that the new lifetime ceiling was no less than 180 days. But never before had a manned vehicle flown a duration not previously flown by an unmanned test mission; there must have been at least a little anxiety over the *Soyuz T-9* health, no matter what the ground tests indicated.

GROWING ANXIETIES

It was this gradually growing age of the *Soyuz T-9*—which brought cosmonauts Lyakhov and Aleksandrov to the *Salyut 7* late in June— that set off alarms and even near-hysteria around the world on October 19. On that day, the *Soyuz* exceeded in age the previously longest *Soyuz* mission. The proximate spark for the panic was a confused and poorly considered BBC report from London. The cosmonauts could not return home safely, nor could they remain aboard the failing *Salyut* space station—a truly cosmic Scylla and Charibdis!

The BBC report was in truth based on authentic anxieties. With the danger of a return to Earth aboard the aging *Soyuz T-9* growing day by day, the cosmonauts may also have been faced with serious doubts about the security of their orbital outpost, the *Salyut*. A report by *Aviation Week* magazine at the time described a serious rocket fuel leak which had allegedly occurred on September 9. Two of the three sets of fuel tanks on the *Salyut* were reportedly involved. The magazine, evidently relying on U.S. intelligence sources who routinely monitor conversations between the cosmonauts and Mission Control in Moscow, claimed that the tanks drained overboard due to a ruptured pipe, and that the cosmonauts were about to evacuate the station in emergency mode before Mission Control advised them it was safe to stay. Since the station had then been pumping a new supply of rocket fuel from the unmanned tanker spacecraft *Progress 17*, a failure in the fuel transfer apparatus was also plausible.

At the time, Soviet spokesmen vehemently denied any such problems. However, at the postmission press conference in Moscow, space officials admitted to Western journalists that a "small" leak had in fact occurred. The failed system was reportedly shut off and, supposedly, there had been no impact on the mission. This confirmation gave added credibility to the rest of *Aviation Week*'s detailed accounts.

Subsequent to the date of the fuel leak, the Soviet cosmonauts did indeed follow a flight plan entirely consistent with such a situation. They greatly restricted the station's pointing maneuvers over ground targets, for photography, while concentrating instead on passively drifting through space. This flight plan was very useful for materials processing experiments which would be hampered by disturbances induced by rocket firings; according to Soviet news reports the cosmonauts conducted exactly those kinds of activities in the weeks immediately following the reported date of the leak.

Another tanker, *Progress 18,* linked up to the *Salyut* on October 22. It conducted refuelling operations a few weeks later, and the cosmonauts reported that all their fuel tanks were "fully topped off." This suggested that the leak problem had been solved and that it may have been associated not with the *Salyut* station itself but with faulty equipment on the *Progress 17* tanker. Support for this theory is the fact that after *Progress 17* was separated from *Salyut* on September 18, it almost immediately plunged back into the atmosphere and burned up. Normally, Progress ships go through two days of additional flight tests before this maneuver. The fact that *Progress 17* skipped this program may indicate that it had problems.

And even if the *Salyut*'s propulsion system failure reoccurs or gets worse in the future, the Soviets still have left one "ace in orbit." This is another unmanned space vehicle, such as *Kosmos 1443*, designed to act as an add-on module to the *Salyut* space station (see Chapter 9).

A RESCUE MISSION?

While the West watched anxiously for some sort of Soviet rescue operation in orbit that October, the path through space of the *Salyut-7* space station made it clear that indeed the Soviets were preparing a new launching.

Any ship sent up to reach the space station can be launched only during the few seconds every day when the Earth's rotation carries the rocket launching pad through the station's orbit plane, a disk which intersects the Earth's center and one continuous ring around the globe. That moment can be computed well in advance if the orbital characteristics of the station are known. Such data is routinely released by NASA, and experienced analysts used it to compute a list of potential launch times, one per day.

Past Soviet practice has been to launch a "chaser" space vehicle—say, a Soyuz or a Progress—when the target Salyut is a standard distance ahead in its ninety-one-minute orbit. For manned launchings, which require one full day to complete the rendezvous, the Salyut is given a lead of fifteen minutes. For unmanned launchings, which then take two days for rendezvous maneuvers, the lead is thirty minutes. At five miles per second, this time difference amounts to a position difference of 4,500 miles, one-sixth of the satellite's entire path around the Earth.

The position of the target in its orbit at the moment when the chaser ship can be launched is also easily computed from public data. And this position clearly showed that an excellent launch window for an unmanned ship existed at 12:59 P.M. Moscow time on October 20, 1983. Fairly confident predictions were made that this would be an unmanned Soyuz craft, with a fresh reentry module to insure the cosmonauts having a safe way home.

But the launch, which did occur precisely at the moment predicted, was instead an unmanned (that part was right!) supply ship, *Progress 18.* It linked up successfully at 2:34 P.M. Moscow time, two days later. A few hours later, as the station crossed North America, radio listener Richard Flagg in Melbourne, Florida, heard the two Russian spacemen aboard the station laughing and joking as they opened letters and gifts just arrived from home. "A happy crew" is how Flagg, an experienced Soviet spacewatcher, put it.

Such supply flights must occur every two months or so to restock the life-support consumables aboard the station and to bring up new equipment as needed. According to Moscow, the ship was also carrying rocket fuel to replenish the *Salyut*'s drained tanks.

A TEST OF *SOYUZ*'S LIFETIME

Meanwhile, the *Soyuz T-9* continued to grow older and older. But under automatic interrogation from the ground (Flagg and his colleagues were picking up intense bursts of radio telemetry data from the *Soyuz,* informing Soviet space engineers of the status and health of all its systems) the *Soyuz* seemed to be receiving a clean bill of health. To verify the operational condition of the *Soyuz* in space, conditions were being compared to those of a test vehicle in a vacuum chamber back on Earth.

After the successful mid-November return of the cosmonauts aboard the unfairly maligned *Soyuz T-9,* some Western sources made it clear that they had known all the time that the ship had never been in any danger since more powerful batteries and fuel tanks had been installed. But this was hindsight; the ship had been in flight for at least a month beyond its planned landing date (and its prelaunch lifetime "redline"). The excess ate deeply into the ship's safety margin, as Ryumin quietly admitted later.

It certainly is going to be of great benefit for the Soviets to qualify their Soyuz ships for missions of such great duration (150 to 180 days instead of 100 days), and thus they may have received a major bonus by being forced to do so this time. Such an extended lifetime would allow a reduction in the required number of "Soyuz-swap" visiting missions, at great savings in money and crew training time. Two full "orbital precession periods" (that is, the progression of landing times earlier and earlier through an entire 24-hour period), plus a standard

Cosmonauts Lyakhov and Aleksandrov (shown on landing), who successfully reached and then nursemaided *Salyut 7* in 1983.

eight-day overlap, come to 124 days. And since Soviet space doctors have said that they believe a 3 to 4 month space duty tour is optimal in terms of health and efficiency, one Soyuz launch every 124 days could carry each sequential relief crew. This would totally eliminate the need for any more brief visiting missions at all. And that could prove to be an unexpectedly advantageous fallout from this late-1983 problem situation.

AN ALTERNATIVE RESCUE PLAN

Concern for the possible hazard to the lives of the cosmonauts led several Western commentators to ask whether or not the Russians could be rescued by the space shuttle. There does happen to be an international treaty requiring all parties to render assistance to space travelers in jeopardy.

In practical operational terms, a space shuttle could indeed reach the Salyut's orbit and approach as closely as necessary for the two men to leap across in spacesuits. But at the moment of greatest danger to the *Soyuz T-9*, neither the *Columbia* nor the *Challenger* could have been readied for launch within many weeks. During that interval, the Soviets would have had plenty of time to prepare and launch their own Soyuz rescue ship—and they have in the past claimed to keep a reserve ship and booster on standby status, as a routine practice, for just such emergencies.

In real-world political terms, such a rescue mission would have to be deemed highly unlikely because Moscow would never have invoked the treaty. In the past, official Soviet policy in the case of emergencies at sea, in the air, or during exploration expeditions in Antarctica, has been to have the endangered men literally rather die than ever be humiliated by asking for Western help. On occasion, Soviet citizens have indeed died rather than be saved by Western would-be rescuers. Cynically, the only situation in which the Soviets might be expected to ask for an American rescue flight is when they know for certain that such a mission could never be mounted in time, and then they could blame the United States for the deaths of the cosmonauts.

THE *SOYUZ T-10A* EXPLOSION

The Western overreaction to the potential (but gravely exaggerated) "marooning" of the *Salyut-7* cosmonauts distracted attention from the growing understanding of the event which had set up the entire crisis, the launchpad explosion of *Soyuz T-10A* late in September.

At the annual conference of the International Astronautical Federation, in Budapest, a top Russian spacecraft engineer (and former cosmonaut) publicly acknowledged the incident and gave further details. According to Dr. Konstantin Feoktistov, a fire had been detected at the base of the rocket at ninety seconds prior to lift off. Ten seconds after the fire had first been noticed, it was confirmed, and the launch director in the blockhouse adjacent to the pad gave the order to fire the *Soyuz* spacecraft's launch escape system.

The men had been within a whisker of death at this point. The rapidly spreading fire at the base of the booster, fed by kerosene gushing from a ruptured tank or pipe, had burned through the electrical umbilical cables over which the on-pad abort command was to have been sent. After this was realized, a radio signal had to be triggered instead. Precious time was spent in discovering the fact of the burned cables, and it took more time to coordinate pushing buttons simultaneously in two separate bunkers. During the seconds of this additional delay, the flames reportedly totally engulfed the rocket and the spacecraft atop it. The two cosmonauts in the *Soyuz T-10A* had no windows to see the flames outside, but they surely could hear—and fear—the excited voices over their radio headsets.

Since the launch window for reaching the *Salyut 7* station had been 11:38 P.M. Moscow time that evening (Monday, September 26), the blast must have occurred about 11:36 P.M. Some Western sources used local launch center time, which at Tyuratam in Central Asia was 1:36 A.M. on Tuesday, September 27. As Soviet manned space events are always dated to the current clock in Mission Control in Moscow, the event can be said to have actually occurred on September 26. It is by this date that cosmonaut Lyakhov, who had been in orbit awaiting his relief crew, referred to the event at the postmission press conference in Moscow. His words, in response to a question from a Western newsman, were never picked up by the Soviet news media, which has kept the Soviet public totally in the dark about the near-tragedy.

The profile of the Soyuz spacecraft's on-pad abort rescue system is a severe one—although of course preferable to experiencing the blast of more than half a million pounds of kerosene and liquid oxygen, which would have created a fireball a thousand feet across lasting ten or twenty seconds. To escape this, a solid-fuel rocket engine mounted in a tower at the tip of the space vehicle fires to pull the cosmonauts clear. The acceleration is intense, more than ten times the force of gravity, sufficient to bring the spacecraft from standing still to the speed of sound in just three seconds. The capsule then falls free, arcing up to an altitude of several thousand feet. There an emergency parachute deploys (the craft is too low to allow time for the primary recovery

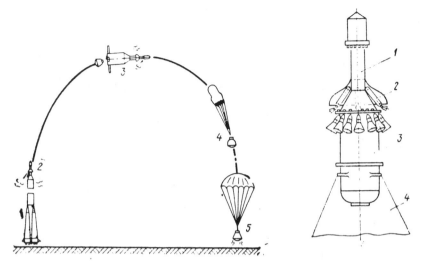

Launch escape sequence. *Left*, (1) booster failure occurs; (2) tower fires, pulls Soyuz clear; (3) command module drops free; (4) emergency parachute deploys; and (5) touchdown. *Right*, (1) steering/jettison motor; (2) steering/jettison nozzles; (3) main separation motor; and (4) spacecraft structure.

Apollo launch escape system test, 1968, probably looked similar to Soyuz firing.

parachutes to deploy), and the cosmonauts touch down very hard about a mile or two from the inferno on the launchpad.

There was initial confusion in the West about who might have been on board, and even if they had survived or not. News of the explosion "leaked" in Washington late on Thursday, three days after the event, and by Saturday had reached the wire services and the *Washington Post*. By then, the best guess was that the crew had indeed survived.

Weeks earlier, an informed source in Europe with close connections to French space officials who frequently visit Moscow had told associates that the next visiting crew (on a Soyuz swap mission) would include another woman-cosmonaut, known only as Irina. Accompanied by veteran cosmonauts Yuri Romanenko and Viktor Savinykh, the woman would visit the *Salyut* and while there take part in a space walk.

This widespread expectation, together with initial garbled reports about a woman being involved in the explosion, led to some published accounts that this Romanenko crew had been the one atop the explosion on September 26. But other sources, both in Washington and in Moscow, soon made it clear that only two men had been involved—Titov and Strekalov. They were privately described by one Soviet official as "healthy but very unhappy."

And well they might have been. Those two men had already been part of one earlier abortive attempt to reach the *Salyut* (see Chapter 8). Now their second try in five months had also failed.

Meanwhile, Serebrov had been removed from the crew to allow the tanking of more rocket propellant since the *Salyut* orbit was higher than in April. A similar reason had forced the removal of Serebrov's backup, reportedly a rookie engineer named Volkov, from the *Soyuz T-9* mission in June. As if to show their good health to those "in the know," Titov and Strekalov, talking by radio linkup to Lyakhov and Aleksandrov, were shown on Moscow television from Red Square during the November 7 parade. Meanwhile, cosmonauts Kizim and Solovyov had been seen on television demonstrating the EVA (extravehicular activity) repair procedures with the *Salyut*'s solar panels.

In the West, analysts were asking the question about the explosion: why now? After more than one thousand launchings of this basic booster (the *semyorka,* in old Russian space slang, or the A-2 in Western civilian analysts' classification), had one of them wiped out one of three available launchpads, at the cost of hundreds of millions of rubles worth of equipment? Over the previous twenty-six years of use of this vehicle, on-pad explosions had been extremely rare.

It might just have been bad luck. But there might also have been some recent tampering with the booster itself, modifications designed to increase its thrust or fuel load, or to reduce its weight. Such modifications, it turns out, are urgently needed because of the extremely tight weight margins imposed on the *Soyuz T* spacecraft by the limited performance capabilities of the *semyorka* booster.

The Soyuz spacecraft (in its modified configuration, dubbed *Soyuz T* (with T for transport) has been requalified to carry three crewmembers, and has done so once in 1981, twice in 1982, once in 1983, and several times in 1984. But in fact it has done so only when the space station's altitude has dropped to below 200 miles. When the station is at its normal operating altitude of 230 miles or higher, the Soyuz simply cannot carry sufficient fuel supplies to bring its full crew complement that high. The weight of one crewman and his equipment, about 220 pounds, must be added in fuel—and subtracted from the crew.

Such appallingly narrow operating margins may have induced Soviet rocket engineers to implement various upgrades to the basic launch vehicle, in order to provide more payload weight in orbit and allow a full three-man crew to reach the Salyut at its nominal operating orbit. That would require an increase in power of only about one percent, or a decrease in structural weight by about the same fraction. But after twenty-six years all the safe and easy improvements have certainly already been made in the *semyorka*. To squeeze anything more out of the rocket was bound to entail risks.

Under these conditions, the September 26, 1983, booster explosion may not have been entirely a total surprise. Nor may it be a freak accident, isolated and absolutely unlikely ever to happen again.

As 1984 opened, the bad luck continued for Strekalov. He and Titov were evidently backup cosmonauts for the *Soyuz T-10* mission which began in February. Normally, they could expect yet another chance to get aboard the space station for a long mission later that year. But only a few weeks later, the 52-year-old Russian flight engineer of the *Soyuz T-11* "Indian visit" mission was medically disqualified. Strekalov was quickly inserted in his place for the week long visit in April.

Past Soviet practice has been to limit cosmonauts to three orbital missions. Strekalov's chance to spend months aboard a space station is probably gone for good.

11

Russia's Space Shuttle Programs

The fleet was far from home and near enemy shores. Seven ships, including a guided missile cruiser and three heavily instrumented tracking vessels, steamed in formation around their assigned duty station. Overhead, aircraft from unfriendly nations kept a close watch; around the fleet, their warships lurked, both on the surface and beneath it. The risk was great, but the mission's importance was greater.

Then, out of a dawn sky, the reason for the unprecedented naval expedition made itself manifest. At supersonic speeds, a streamlined delta-winged craft soared into view. Instruments aboard the tracking ships testified to the vehicle's recent plunge back into the atmosphere from space, at Mach 25. Under automatic guidance (for nobody was on board), it had survived the fiery reentry and now was nearing its target in the southeastern Indian Ocean.

A sonic boom echoed over the waves just as the small spaceplane popped a parachute for the final descent. With a splash and the hiss of seawater encountering a still-glowing heat shield, the mission ended.

Frogmen jumped from a helicopter and approached the floating capsule. On the spaceplane's side was written an identification of its point of terrestrial origin: CCCP. The Russian frogmen read it as "SSSR" and recognized it as the abbreviation for the Union of Soviet Socialist Republics, their distant homeland. High overhead, the Australian patrol planes photographed the operations.

Within hours, the Australians' film was rushed to military intelligence analysts in the United States. After years of myths and rumors, hard evidence was finally at hand. It was June 1982, and the homecoming of the Soviet Union's *Kosmos 1374* had just been observed and recorded. Russia's "space shuttle" was real.

Or was it? Things usually not being what they seem in the enigmatic and obscure world of Soviet space technology, nothing so obvious could possibly be so true.

For at least a decade, Western observers had been awaiting the advent of the Soviet space shuttle, Russia's technological riposte to the NASA space fleet of the *Enterprise, Columbia, Challenger,* and their sister spaceships. From Soviet spokesmen came conflicting and enigmatic hints. Intelligence sources spoke of strange preparations at secret Soviet spaceports. A project named Albatros was said to be a fully reusable spaceplane which would catapult Soviet cosmonautics a generation ahead of would-be competitors in the West. At the time of the drop tests of *Enterprise* in mid-1977, reports in Europe claimed that the Soviets had already completed their equivalent program and were racing toward an orbital launching. The late 1970s saw the construction of a giant new runway at the Tyuratam space center, a landing strip so huge that it was clearly visible in low-resolution photographs taken by NASA's *Landsat* earth resources survey satellite. Beginning in late 1976, baffling and unexplained space vehicles had been shot into space on quick "once around" hops, obviously to test something new—but what? Later, testimony by high-level American defense officials began to spell out the existence of a Soviet program allegedly similar to the USAF's *Dyna Soar* (or X-20) space glider of the early 1960s, cancelled before first flight.

Surely here was a puzzle to warm the heart (and try the patience) of any analyst. Moreso even than Russia itself, the Russian space shuttle was (to abuse the Churchillian metaphor) an enigma, inside a mystery, wrapped inside a puzzle.

SOVIET DISCLAIMERS

Official Soviet statements had never been of much help in resolving the issue. In recent years, they seemed generally to indicate that the

USSR was not really interested in such a project, at least in the near future.

June 1980: Chief cosmonaut Lt. Gen. Vladimir Shatalov, three-time space veteran, told a Hungarian journalist that "Soviet specialists have also investigated the possibility of producing spacecraft which can be used more than once. In this given stage, however, they consider that the employment of these spacecraft is not justified for Soviet research because the present tasks can be solved with the well-tested methods in an economic way."

That same year, cosmonaut Georgiy Beregovoy, mayor of "Starry Town" where Soviet spacemen live and work, wrote that "with years some changes will certainly be made to the orbital station and transport ships and new instruments will appear aboard them, but the present pattern [Salyut, Soyuz, Progress] of using the station and the basic elements of this orbital permanent system will probably remain unchanged for a long time to come."

April 1981: Cosmonaut Vitaliy Sevastyanov, who hosts a popular Soviet television program on scientific topics, declared, "We are also thinking—on a long-range basis—of a reusable space shuttle. However, at present detailed calculations confirm that the use of our existing system of transport of persons and cargo, as well as orbital stations of the new generation, will be much cheaper than with a space shuttle—in the next decade at least. We will build large modular orbital stations suitable for a longer stay of dozens of specialists, and for this purpose our transport techniques are more suitable." A few months earlier, at a space conference in Paris, Sevastyanov had said that he expected the Soyuz-Salyut-Progress system to remain in use for the next fifteen years.

Later in 1981, veteran cosmonaut and Mission Control Center director Dr. Aleksey Yeliseyev was asked if the USSR was interested in a space shuttle of its own: "As in any sphere of engineering, the quest for new solutions is conducted in diverse directions, multiple-use technology being one of them," he answered. "But at present we are quite satisfied with our new *Soyuz T* series."

April 1982: Dr. Yuri Zaitsev, a department chief at the Space Exploration Institute (the Soviet NASA), wrote that "the supply system for long-term orbital stations, based on the relatively cheap Soyuz manned spaceships and Progress automatic supply vehicles, should be considered most efficient and advisable today."

Earlier that year, a Soviet science attaché in Washington had told a luncheon meeting of American space experts that a Soviet space shuttle was a "very complicated effort that will take much time and money." Anatoliy Skripko thought that development might begin in five years or so, but that there is presently no urgency since "it is no problem for us now to deliver fuel, food, and other supplies" to space stations using currently available space vehicles.

PROJECT ALBATROS

The Soviet disclaimers contrasted sharply with the flurries of rumors which circulated in American and European space circles in the late 1970s. Most spectacular of the stories was the one about Project Albatros (the Soviet spelling of albatross), which supposedly consisted of a fully reusable two-stage system of winged space vehicles. The craft was to be launched from a speeding hydrofoil on the Volga or the Caspian sea; the first stage, much bigger than a C-5A transport, would carry the Orbiter to the edge of space, and then glide back to a runway; the Orbiter itself would continue into space for its cargo mission before returning to the same runway a few hours or days later. The Soviet terminology for such a vehicle reportedly was *raketa-plan* (rocket-plane) or *kosmo-lyot* (space-flier).

Most serious observers considered this project far too advanced for a "first step" into space-shuttling operations. The Soviets had demonstrated neither the materials, the space-borne computer capacity, nor the high-energy rocket engines necessary for such a system. And this skepticism was borne out later when the true origin of the Albatros project became known: it was merely a detailed design study by students at a Moscow aeronautical engineering institute. The real Soviet manned space program evidently had no connection.

MYSTERY FLIGHTS

Something new, however, was definitely being tested at the Tyuratam spaceport. Between 1976 and 1979, four giant Proton rockets climbed into space, each carrying a strange cargo of twin unmanned spaceships. The official Moscow announcement merely described them as routine scientific satellites. But Western observers were able to determine far more information—and all of it was baffling.

The launchings were all at night, "on the hour." Once in orbit, two different payloads separated from the giant booster. The payloads circled the Earth once (on one occasion, twice) and then fired retrorockets to plunge back into the atmosphere. The spacecraft each weighed about ten tons. Landing occurred in the standard Soviet recovery zone in Central Asia just after dawn. The recoveries, according to Western intelligence agency sources, were all successful, and the payloads maneuvered during descent, just like winged space shuttles would.

It was easy to describe these flights as space shuttle tests, and Western newsmen did so. But then in 1979 the test flights abruptly ceased. Experts speculated that the flights might have been merely aerodynamic tests. Perhaps the program was a failure—the flights could have shown that such a vehicle was not feasible at the current level of Soviet space technology.

These mystery flights—which to the present day have not been satisfactorily explained—coincided with some boldly explicit Soviet statements about their "shuttle" plans. Unlike the pessimistic and conservative comments which were shortly to follow, these declarations were upbeat. Cosmonaut Beregovoy had told German newsmen in 1979 that a space shuttle was "a logical next step" and that "we don't want to be left behind." Cosmonaut Anatoliy Filipchenko revealed: "We are trying to build a vehicle we don't have to throw away." Two-time space veteran Pavel Popovich teased an East German reporter with obscure hints about having such a vehicle in orbit prior to the launch of the American counterpart, and fellow-spacemen Georgiy Shonin and Yevgeniy Khrunov echoed those sentiments that same year. These comments could not have been mere idle boasts—Soviet cosmonauts never ad lib, and all their public statements are cleared in advance.

A MINI SPACE SHUTTLE?

But 1980 heard a different tune, as already described. Now the unanimity had swung 180 degrees—the USSR was *not* interested in building a shuttle soon. Something must have happened to change the minds of Soviet space experts.

Early that same year, the U.S. Secretary of Defense, Dr. Harold Brown, told a congressional committee that he had seen indications that the Soviets really were building a space shuttle. "There is evidence that they are working on something," he said. "It probably is more along the lines of the earlier U.S. program, the DynaSoar, than something of the capability of the Shuttle." NASA Administrator Robert Frosch confirmed this suggestion. And in 1981, Dr. Robert S. Cooper, director of the Defense Advanced Research Projects Agency (DARPA), publicly announced that the Soviets were "working on a manned space plane."

Against the background of the mystery test flights (ending in 1979), the subsequent Soviet switch to denials of any immediate interest, and the American hints that a spaceplane was under development, the June 1982 Indian Ocean splashdown came as quite a shock to most Western observers—who had generally expected something a little more advanced and somewhat bigger. It (and nearly identical follow-up flights in March and December 1983) was quickly interpreted as the expected (but denied) spaceplane—but there were several glaring loose ends.

All three launches had been from the small Kapustin Yar rocket base, on the lower Volga River. The first two launches had been at about midnight. The boosters, modified medium-range missiles similar to the American *Thor,* had carried the payloads southeastward across the Aral Sea and the Himalayas and into very low orbits about 100 miles up. The one-ton payload separated from its booster rocket (which stayed in orbit several days before burning up) and circled Earth once, crossing Malaya, Java, Australia, the South Pacific, Panama, the North Atlantic, England, Central Europe, and then the Crimea. Over the Crimea, where the main Soviet space-tracking facilities are located, the satellite turned tail forward and fired a small rocket engine to initiate its return to Earth. It fell back across Iran, southern India, and Sri Lanka, where it filled the skies with a fiery streak. Minutes later it landed, south of the Cocos Islands in the eastern Indian Ocean, amidst the waiting Soviet recovery fleet and the uninvited Australian snoopers.

The December 1983 test (*Kosmos 1517*) made its deorbit burn

Soviet spaceplane being recovered by frogmen after splashdown in Indian Ocean in 1983. Vehicle appeared to be a subscale model weighing about one ton. (Photograph courtesy of Australian Defence Ministry).

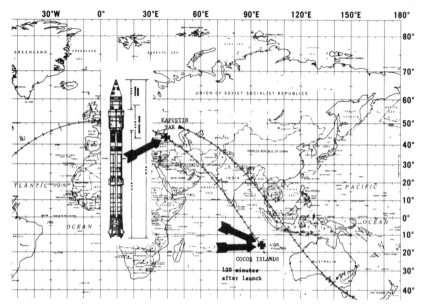

Flight path of spaceplane tests shows strange features. Two vehicles *(twin arrows)* splashed down near Cocos Islands; the third came down in the Black Sea *(single arrow)*. Booster used has no operational pads at Tyuratam, probably explaining use of Kapustin Yar range.

earlier and plopped down in the Black Sea—a major improvement. And this time Moscow officially admitted the recovery.

The photographs from the first of the three landings showed a streamlined craft with one stubby vertical stabilizer and two outboard aft stabilizers, both tilted outwards. The front of the fuselage showed what for all intents and purposes looked exactly like a three-paned cockpit window. From a top hatch amidships, parachute lines extended and a long conical inflated balloon protruded—presumably some sort of recovery aid. The whole craft could not have been more than twelve feet long and perhaps eight feet across, from wingtip to wingtip (that would have just fit under the standard nose cone used atop the booster rocket).

"WEAPONS TESTING"

The "mini-space-shuttle" explanation was so obvious and so neat that some skeptics resisted falling for it. A few private analysts even developed the heretical notion that these tests really had nothing to do with the Soviet manned space program at all. Instead, these observers suggested, the winged objects fished out of the Indian Ocean were dummy thermonuclear warheads designed to drop out of orbit and attack Western naval task forces.

Their logic is seductively straightforward. The launch site, Kapustin Yar, has never before hosted any man-related missions—but it has been involved in weapons testing for thirty-five years. Only two earlier Soviet space vehicles ever splashed down at sea, and that was only because they had come back from the Moon and were lucky to reach anywhere on Earth—a splashdown would never be a feature deliberately chosen for a man-related vehicle, the heretics argued. If the flights were merely aerodynamic tests, there would have been no need to go all the way into orbit since a simple suborbital mission with extra acceleration could easily have matched the desired velocities. The orbital mission required the development of a special "service module" complete with power source and retrorocket—a waste if the spacecraft were merely a subscale model. The shape of the vehicle is

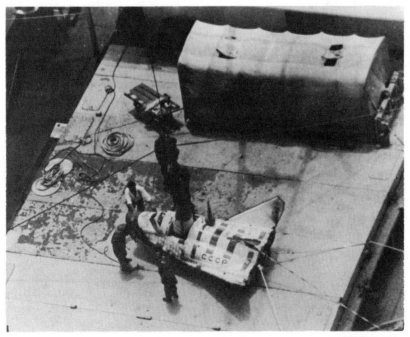

Soviet spaceplane on deck of recovery ship, the *Yamal*. (Photograph courtesy of Australian Defence Ministry.)

no puzzle at all since any antifleet warhead must have such a winged shape in order to perform "crossrange maneuvers" needed to get to the target from a nearby orbit.

Would the Soviets build such a weapons system to attack American aircraft carrier task forces, for example? The heretics argued that they would since the USSR had already developed a nuclear-powered ocean surveillance radar satellite (like the kind which has three times fallen out of the sky at random, most recently in January 1983) precisely to locate such fleets—and there are likely to be American fleets near enough to be dangerous but at the same time out of reach of other Soviet antifleet weapons. Going into orbit and coming back does not exact a significant payload penalty and it does insure a standardized descent profile—something that the use of ICBMs would not do. Any aiming inaccuracies can be taken out with terminal guidance. And despite the fact that such a weapon is outlawed by international treaty, the Soviets have already demonstrated their regard for such a treaty by developing (in the 1967–71 time period) the so-called Fractional Orbit Bombardment System (FOBS) intended for placing nuclear warheads into temporary low orbit for surprise attack.

The 2,000-pound winged vehicles would be plenty large enough to carry a warhead of several hundred kilotons, continue the heretics. The launch vehicle is small enough to be based in already-existing missile silos throughout the USSR. And the "windows" on the nose of the vehicle are needed for the terminal guidance sensors to see through, not for cosmonauts' eyes. No landing system is needed since the vehicle will detonate high in the air.

Superficially, this terrifying interpretation does seem to explain more of the factors of these missions than does the "obvious" view of them as space shuttle tests. But to everybody else, the vehicles "look like space shuttles"—and presumably (and hopefully!), government analysts have additional classified data upon which to base their judgments that the weapons idea is "fantasy."

A SOVIET MANNED SPACEPLANE

Two days after the second splashdown, DARPA director Robert Cooper unequivocally identified the Soviet vehicles as development tests for a Soviet manned spaceplane. But it still remained obscure how much the test model would be scaled up for the manned version. Various estimates ranged from a factor of two to a factor of four in all dimensions.

Different attempts were made by U.S. analysts to compare the

Indian Ocean splashdown vehicles to equivalent American craft. In the 1960s, U.S. space engineers ran several reentry programs to investigate aerodynamic conditions at very high velocities. The programs carried names such as ASSET (Aerothermodynamic/elastic Structural Environment Test) and PRIME. The latter involved launching subscale models of the X-24 *lifting body* on Atlas boosters from Vandenberg AFB in California to a recovery zone near Kwajalein Atoll in the mid-Pacific. In the late 1960s there were manned low-speed lifting body flights at Edwards AFB, involving such vehicles as the X-24, the M2-F2, and the HL-10. The latter reportedly had the best flying characteristics, and it is the U.S. aerospace vehicle which the Soviet test spaceplanes most closely resemble.

A GIANT SHUTTLE

The second Soviet mini-spaceplane flight in March 1983 could not have been better timed to highlight one section of a Pentagon special report. The document, released by the Office of the Secretary of Defense, was entitled "Soviet Military Power"—and in a special section on space vehicles, the report gave uncharacteristically specific parameters on not one but *two* distinct Soviet space shuttle projects.

The first project was just the already-known spaceplane concept: "Orbital development test flights of the smaller vehicle have already

Early NASA design for an eight-man lifting body (with cargo module to the right) could approximate what the Soviets might build twenty years later, in the mid-1980s. (Photograph courtesy of NASA.)

occurred,'' the report claimed. But the second project allegedly involved a spacecraft with much greater performance capability than the NASA space shuttle.

The Pentagon report described a vehicle disturbingly similar to the U.S. design but with both smaller lift-off weight and thrust and a larger payload into orbit. Like the U.S. system, the Soviet system is to consist of a winged Orbiter, a large disposable External Tank (half again as long as the U.S. equivalent, however), and two strap-on booster units for lift-off assist (the Soviet system evidently will use liquid-fuelled boosters, unlike the solid-fuelled ones employed by NASA). The Soviet shuttle "could be in regular use within a decade," the report claimed.

This information supposedly came from actual observations of existing test hardware. During the week the Pentagon report was being prepared for release, the Washington newsletter *Aerospace Daily* obtained a more precise description of the Soviet space shuttle Orbiter vehicle, based on spy satellite photographs of the craft at the Ramenskoye flight test center near Moscow. The Soviet spaceship was 109 feet in length (NASA's is 122 feet) with a wingspan of 76 feet (NASA's has 78); the Soviet fuselage diameter is 18 feet and the wing leading edge sweep is 46 degrees—little different from the *Columbia* or the *Challenger*. Both the Soviet Orbiter and the giant External Tank are transported in cradles atop modified Bison bombers—and supposedly that is why the giant new runway at Tyuratam had been built several years ago.

A few weeks later, even better photographs were reportedly obtained when one of the Soviet Bison carrier aircraft, with an Orbiter mounted atop, reportedly ran off the runway at Ramenskoye and got stuck in the mud for two days.

Such observations allowed U.S. analysts to obtain information on the design differences between the U.S. and Soviet vehicles. For example, the Soviets evidently are still unable to build powerful and efficient rocket engines such as those which propel the U.S. vehicle. Therefore they are sticking to expendable engines, mounted on their large External Tank (which then by definition should be counted as a rocket stage and not merely a tank). Without engines at the aft end of the Orbiter, the Soviets can streamline it (engineers call it a "boattail") and give it twice the aerodynamic handling capability of the U.S. system. Additionally, if the Soviets chose to build their payload doors at the aft end, like clamshells, they could save even more structural weight and complexity over the U.S. approach.

But without reusability of engines and boosters, the economy of such a system remains dubious. Just perhaps, economy is not what the

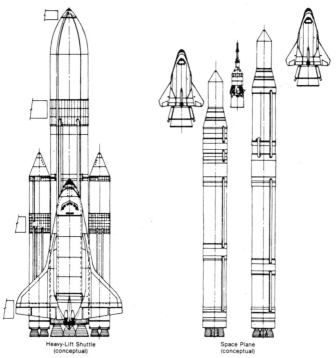

Heavy-Lift Shuttle
(conceptual)

Space Plane
(conceptual)

Possible configuration of full-size manned spaceplane, aboard new intermediate launch vehicle, compared to Soviet heavy-lift shuttle. (Illustration courtesy of C. P. Vick © 1983.)

Fanciful Soviet painting of a two-stage fully reusable shuttle for the next century.

Soviets are after. As Gen. James Abrahamson admitted in mid-1983, NASA's design never delivered the launch economy which had been anticipated, but it did provide something almost as good: "special services." Speaking to a convention of the Aviation/Space Writers Association, Abrahamson, NASA's associate administrator for space flight and a former USAF Manned Orbiting Laboratory astronaut, explained that "that service, being able to provide a man or a woman in orbit to tend equipment and to exploit what happens there, may be the most important service the Shuttle provides for the future." He explicitly pointed to the development of the Soviet shuttle as proof that they too are convinced of the value of such special services.

SEARCHING FOR ANSWERS

If an observer might have expected that this flurry of new data on "Soviet space shuttle programs," which burst forth onto the public record early in 1983, would settle any of the mysteries, then that observer must not have been paying attention. More data there certainly was—but did it provide any more answers?

At least one intriguing pattern has emerged. In the late 1970s, the Soviets secretly conducted expensive "landing tests" of some sort of advanced spacecraft; during that same period, cosmonauts made confident predictions about the imminent advent of a Soviet space shuttle. The tests stopped in 1979; the cosmonauts' public statements reversed completely in 1980, with the new line being that the Soviet manned space program had no immediate need for a space shuttle. An observer may suspect a cause-and-effect relationship between those events. Something, evidently, had *not* gone well at the space testing range.

The first hard evidence for a Soviet shuttle appeared in mid-1982 with the first Indian Ocean test. If anything, it was a step backward from the late 1970s tests and reflected a technology mastered in the mid-1960s by American space engineers. A few heretics have hypothesized that the vehicles are not even man-related at all—but even if they are, it could be several years before a full-scale manned version is ready for testing. At that rate, such a vehicle could hardly become operational by the end of this decade. That fits in very nicely with the Soviet decision to spend a great deal of money developing the *Soyuz T,* a totally rebuilt manned space vehicle which is justified only if its developmental costs are amortized over many years of future operation. It is difficult to imagine any manned orbital missions which a spaceplane can conduct in 1990 that the *Soyuz-T* cannot already conduct in the mid-1980s.

As for the "giant shuttle" which the Pentagon describes (and which "leaks" of spy satellite imagery descriptions support), its purpose is even more enigmatic. Since its boosters and engines are supposed to be not reusable, economy is difficult to postulate as a justification for it.

Perhaps the answer can be found in the propaganda blitz which Moscow has unleashed against the American space shuttle program. It is not for economy either, they claim, but for vicious military purposes. Soviet news broadcasts consistently portray the NASA spaceships as being weapons carriers (including H-bombs), weapons platforms, satellite interceptors, and "space pirates" out to kidnap peaceful Soviet scientific satellites.

Conceivably, Soviet leaders believe such nonsense, or Soviet military planners fear it. They then would decide, whatever the cost, to create their own corresponding system as a counterbalance.

Why then are the Soviets building such vehicles (if indeed they really are)? Perhaps it is merely because the United States has them. If so, a potentially wasteful technological detour of immense difficulty and cost, and of uncertain success, has been brought on by Soviet paranoia and gross misperceptions of U.S. intentions and capabilities in space. If so, this project could detract and damage any of the really valuable Soviet programs, such as their permanent space station—and it could help bring about in reality what the Soviets only currently imagine.

Perhaps that mythical code name, Albatros, wasn't really so far off base! The Soviet space shuttle, in whatever manifestation it is finally incarnated, does have a touch of the fantastic and even the whimsical about it. Observers hope it will remain as harmless.

12

A Shuttle-Salyut
Joint Mission

Sometime in 1985, an American shuttle-Spacelab mission and a Soviet Salyut space station complex will be in orbits so close together that the astronauts and cosmonauts aboard would need do little more than reach out to shake hands with one another.

Such a space handclasp would be far more productive than the exciting but largely symbolic *Apollo-Soyuz* linkup in 1975. But in the mid-1980s, with U.S.-USSR tensions at high levels, the symbolism of space cooperation is nothing to sneeze at either.

A Shuttle-Salyut program was once generally regarded as a logical follow-on to ASTP, or the Apollo-Soyuz Test Project (Soyuz-Apollo, in Soviet space history books). But the rumbling disintegration of detente in the late 1970s (coinciding with the very real and very loud rumblings of Soviet tanks in Saigon, in several African nations, and ultimately in Afghanistan) forestalled any real progress, and there have been no official new negotiations in many years. A few years ago, the Reagan administration called a halt to remaining bilateral U.S.-USSR space projects (while leaving multilateral agreements in force—a loophole we may be able to launch a spaceship through). Politically, the idea of a new space linkup is absurd—at least at first glance.

A LOGICAL MOVE, TECHNOLOGICALLY

In the meantime, the technological logic for joint manned programs has become even more compelling while political and diplomatic realities have grounded serious speculation. Even nonserious speculation has been banished—the topic is about as taboo as anything can be these days.

However, the logic remains straightforward. Because of different requirements for space operational technology, the United States and the Soviet Union have recently been following different courses in manned spaceflight. The Russians, bedeviled by electronics failures that severely limit the lifetimes of their unmanned systems, have set their sights on permanently man-tended Earth-orbit space stations, served by a space transportation system based on tried and true expendable launch vehicles. Technologically, this is nothing the United States could not have done in the late 1970s if the Apollo/Skylab program had been expanded and continued—which it wasn't.

The United States, seeking a quantum jump in the ease and economy of access to orbit, instead tackled the much more challenging task of building a reusable space shuttle and its auxiliary equipment, such as the Spacelab scientific module and a family of upper stages.

Near the end of the 1980s, both spacefaring nations' paths will probably again converge, as the USSR seeks greater flexibility and economy in transportation (necessitating its own shuttle now under development—see Chapter 11) while the United States establishes its own permanent manned orbital presence. At that point, most of the mutual need for cooperation will go away.

But for the next five years or so, the shuttle and the Salyut will be the two complementary halves of what could be transformed into a complete international manned space effort. It only requires that the appropriate political decisions be made in Washington, Moscow, and the capitals of the lesser spacefaring powers. Only that!! And it must be done soon before the opportunity—and the need—evaporates.

Here's one view of how it might actually happen, given such a decision.

A SPACE-HANDSHAKE SCENARIO

In 1985, several Spacelab missions are to be flown in high-inclination orbits while the Soviets will be operating at least one manned space station in a similar orbit. The Soviet outpost—perhaps *Salyut 7* or even *Kosmograd 1*—will be orbiting Earth at an altitude of 230 miles and an orbital inclination of 52 degrees. Current NASA plans for *Spacelab 2*, meanwhile, call for an orbit of 233 miles high at an inclination of 50 degrees.

The *Spacelab 2* mission is an astrophysical expedition with pallets full of instruments but no pressurized laboratory module like *Spacelab 1*. The shuttle could spend a week in orbit with seven astronauts on board. The crewmembers will work in two shifts. The flight crew will be: commander, Gordon Fullerton; pilot, David Griggs; flight engineer, Dr. Story Musgrave. In addition, there will be two NASA mission specialists, Tony England and Karl Henize, and two payload specialists selected from a group of four civilian astrophysicists already in training. The crew will already be trained for projects already approved, to conduct a rendezvous and robot arm activity.

The Soviet station could have as many as six cosmonauts aboard, working on orbital duty tours of three- or four-months duration. The complex will consist of a Salyut-class care module plus a number of add-on specialized modules similar to the *Kosmos 1443* module tested by cosmonauts on *Salyut 7* in mid-1983.

To conduct a space rendezvous between those two missions would not be difficult. The *Spacelab 2* orbit would have to be raised in inclination just a bit, which is merely a matter of altering its launch azimuth from the Kennedy Space Center from 47 degrees to a barely more northerly 45 degrees. There would also be some changes in the mission activities caused by the slightly different path and by the times allocated to carrying out joint activities. But the potential payoff should more than make up for the sacrifice.

Apollo-Soyuz in 1975 was a "proof of concept" of international rendezvous and docking. Shuttle-Salyut, as it matures, can go far beyond mere symbolism. The first mission (and 1985 is not too soon to hope for it) should probably be kept simple and should not involve a physical linkup—but it could demonstrate emergency communications channels, EVA equipment compatibility, and other rescue procedures. A ship-to-ship spacewalk could accommodate ceremonial needs and have the practical value of simulating an emergency evacuation. Small packets of equipment could be transferred to the Salyut for long-term operation in space, after which the results would be turned over to American specialists: the Salyut crew, in turn, could send back to Earth aboard the shuttle recent experimental results.

The equipment for a physical linkup would require some modifications to current hardware but would be nowhere near as complex as the Docking Module carried by the *Soyuz*-bound *Apollo* in 1975. To make things simple, both nations' spacecraft now have the same composition/pressurization of their cabin atmospheres—unlike in 1975. A special tunnel is installed on pressurized-module Spacelab missions, and it has an airlock set atop it in the payload bay. A standard Soviet-compatible docking probe would merely have to be installed atop that airlock, and structural reinforcements added to transfer mechanical docking stresses into the walls of the shuttle's payload bay. Those kinds of modifications could be designed, verified, and built within twelve to eighteen months, once the go-ahead was given.

JOINT SPACE MEDICINE ACTIVITIES

One especially rewarding activity on joint visits could be carried out in the arena of space medicine. In this field, even today there exists a vigorous and mutually beneficial exchange of data worldwide. On the first joint mission, an American astronaut—perhaps even a physician—could visit the Salyut and spend several hours conducting medical examinations to calibrate U.S.-developed space medical-monitoring equipment and to gather additional "data points" for studies of space sickness. Such studies, conducted by astronaut-physicians aboard shuttle missions in June and August of 1983 (see Chapter 6) provided invaluable physiological data. The data collected on Shuttle-Salyut would provide a different baseline, useful for future U.S. long-duration space missions. At the same time, it could give Soviet scientists a different angle on their own space medical observations.

In later missions, which might be repeated at least once annually for the rest of the decade, personnel could be exchanged for longer and longer periods. American astronaut specialists could conduct materials-processing experiments aboard the Salyut's furnaces, perhaps powered by hookups from the shuttle's fuel cells. Soviet cosmonaut-scientists could make scientific observations with American astronomical instruments. U.S.-developed animal cages containing large animals—such as monkeys—could be placed aboard the Salyut for retrieval up to a year later. Ultimately, by the late 1980s, the swapped personnel might even be able to complete their missions and return to Earth aboard the other side's spacecraft.

TRAINING PROCEDURES

The crew-training procedures for foreign guests have already been established in both countries. Ten foreign cosmonauts have already flown in the USSR's spacecraft, with as little as ten months' training. Two European astronauts have been trained in Houston, and another at the Marshall Space Center in Huntsville, Alabama, for European-funded Spacelab missions. For payload specialist astronauts without any direct flight duty, NASA has developed a minimal training program that can be completed in as little as three weeks. The actual preparation for specific productive work in orbit, of course, would take much longer.

PREPARATORY STEPS

Before an actual joint mission, each nation's nationals probably should fly at least a few times on the other nation's spacecraft just for orientation and familiarization. Incredible as it may sound, such flights could easily occur within a year of a formal agreement between Washington and Moscow. Since there are "passenger seats" open on both Soyuz/Salyut supply flights and on routine space shuttle satellite deployments, no additional costs would be incurred by such a decision.

There are several such preparatory steps which could be taken prior to an actual linkup or rendezvous. Emergency landing zones for each other's space vehicles could be designated in each country (the Soviets have a runway at their launch site plenty long enough to handle a returning Orbiter if only some guidance equipment were installed prior to need). Standardized emergency radio frequencies and communications protocols could be set up and appropriate documentation carried on all manned space missions. Permanently manned liaison offices could be opened adjacent to each nation's mission control centers, and hot lines could be set up and tested—before they are needed. Standardized spacesuit oxygen couplings—or at least, ready-made adaptors—could prepare the way for ship-to-ship transfers for routine visits and for lifesaving rescues. Eventually, provisions for ship-to-ship electrical power transfers (for routine augmentation or even emergency "jump starts") could be added.

VALUE OF A JOINT MISSION

Meanwhile, the Shuttle-Salyut linkup has its own compelling logic. The Russians can do what Americans cannot currently do: keep manned platforms operating efficiently in orbit for months at a time. The Americans can do what the Russians currently cannot do: transport large bulks and large crews rapidly and easily from Earth to orbit and back again.

As far as the practical value of such a joint project, there should

Collage shows true-to-scale view of a Shuttle/Spacelab vehicle alongside a Salyut. (Photograph courtesy of NASA.)

Fanciful collage *(by author)* shows Salyut docked to special module attached to shuttle's airlock in payload bay (Photograph courtesy of NASA.)

DOCKING MODULE
MOUNTED ON TUNNEL ADAPTER

DOCKING
MODULE

AIRLOCK   TUNNEL ADAPTER

NASA sketch shows installation of generic "docking module" for linkup with another Orbiter, with a space station, or with a Soviet space vehicle.

be dozens of candidate experiments—if space engineers only think hard about the unexpected opportunity. To deny the utility of visiting and working aboard the world's only (so far) permanent space station is, in effect, to pull the rug out from under arguments that America needs a similar facility. Put another way, if American space scientists cannot think of anything useful to be done by guests aboard a Salyut, then they will be hard put to justify the utility of the proposed NASA space platform. Conversely, when space experts produce a list of desirable experiments which can be done only aboard the Salyut (thus forcing us to go to the Soviets for help), then the argument that "we need a comparable facility" will be immeasurably strengthened.

THE MULTILATERAL ANGLE

Interestingly enough, the United States government need not be the initiator of such a joint space project. If the Reagan freeze on bilateral cooperation remains in force, then the project can be approached from a multilateral angle, with the European Space Agency (ESA) taking the initiative. The key is *Spacelab D-1*, and another newly rescheduled mission called *Spacelab EOM-1*.

In late 1985, NASA has scheduled the *Spacelab D-1* mission. The laboratory module is conducting materials-processing experiments for the West German government and for the European Space Agency, and ESA astronaut Wubbo Ockles will be accompanied by one German scientist (either Reinhard Furrer or Ernst Messerschmid), one or two other NASA mission specialists Guy Bluford and Bonnie Dunbar, pilot Steve Nagel, and two NASA pilot-astronauts as yet unnamed.

The foreign clients are paying cash for the mission, and they specify the flight profile. If the Europeans wish to sponsor a space rendezvous with the Soviets, NASA would have the right to charge them extra for "special services." Besides, all the necessary hardware (such as rendezvous radar dish) and software (such as rendezvous-targetting algorithms in the flight computers) are already paid for and installed. Several hundred extra pounds of maneuvering propellant would have to be provided, along with a few more months of crew training. Those impacts are not negligible, but they can be handled.

The orbit of *Spacelab D-1* is not quite appropriate for a rendezvous. The baseline plan is for a 57-degree inclination, at 202 statute miles. That would have to be changed to 52 degrees, at about the same altitude. The lower inclination would not cover all of German territory, and there is some political pressure to "show the flag" in orbit over the country paying the space freight bill. However, there is only one experiment which is really supposed to need that inclination; it involves tracking sites in Germany which would have to be relocated to the southern part of Europe if the orbit were altered. But such sacrifices may well be worth the spectacular gain.

If *Spacelab D-1* is not to be the precedent-maker, other candidate missions exist. One is the *Spacelab EOM-1* (Environmental-Observation-Mission, a short pressurized module plus one pallet loaded with cameras), a flight designed to conduct many of the experiments missed on *Spacelab 1* in December 1983. It is scheduled for mid-1985 and is not heavily loaded with activities—a *Salyut* rendezvous would be easiest to add here. Since European scientists are deeply involved in this flight, the multilateral nature of the Shuttle-Salyut mission would still exist. The next high-inclination shuttle flight will not be until *Spacelab D-4* in March 1987, two years later. *Spacelab 8* is also scheduled that year, along with a shuttle mission carrying a satellite called *ROSAT*.

COST

The cost of the 1975 ASTP linkup, in today's currency, was upwards of half a billion dollars. Similar expenditures in today's prevailing budget deficits are naturally out of the question. But there is a major difference this time: Shuttle-Salyut would be an add-on mission for spacecraft already scheduled for launch and already paid for. The missions would not, as in 1975, require the purchase of dedicated space vehicles and boosters. Under such conditions it is difficult to imagine how each Shuttle-Salyut mission could cost even one tenth as much as *Apollo-Soyuz* did.

THE OFFICIAL VIEW

NASA's current official attitude about the possibility of such a mission is apparently undefined. After ASTP, follow-on manned missions were seriously considered. But the Carter administration came to discourage such studies, and they quietly ground to a halt.

In late 1978 an internal NASA memorandum was issued to instruct public affairs officers in how to answer public queries about possible future manned cooperation: "A 1977 agreement between NASA and the Soviety Academy of Sciences calls for a study of the objectives, feasibility, and means of carrying out a joint experimental program involving the U.S. Shuttle and the Soviet Salyut," the official NASA statement read. "Preliminary discussions on this subject were held in Moscow last year [1977]. No further discussions have been scheduled pending a comprehensive U.S. interagency review of the entire subject.

We cannot predict when this review will be completed [it never was!] or what the outcome will be." The discussions, in other words, were bureaucratically scuttled.

Moscow got the message. On each anniversary of the July 15, 1975, ASTP launching, Soviet authorities issued statements praising the mission as proof of the value of cooperation while criticizing the Carter administration for cutting off discussions for a next step. Russian cosmonauts regularly told newsmen how sorry they were that no new cooperative ventures were being undertaken.

Back in Washington, meanwhile, the suspended negotiations were evidently soon forgotten. After the election of Ronald Reagan, James M. Beggs became the sixth NASA administrator and was sworn in the following summer. During an interview with writer Trudy Bell, he was asked, "Do you think that there will be any opportunities for opening up joint Shuttle-Salyut missions, maybe when Spacelab gets operational in 1984 or 1985?"

Beggs replied, "The Russians have made absolutely no overtures to us to open up any joint missions in any of the discussions that I have been privy to, or have been informed of. And I don't expect them to, because their perception is that they have moved out ahead of us for the moment." The 1977 official Shuttle-Salyut bilateral agreement (signed by both parties) evidently had fallen through the crack of administration transition, the victim of institutional amnesia.

One major public perception that could seriously hinder such future missions is the widespread popular attitude in the United States that the 1975 *Apollo-Soyuz* was some sort of "space rip-off." In it, the Russians were supposed to have stolen American space secrets while exploiting a propaganda bonanza by unfairly posing as an "equal space partner" with an American side which actually did most of the flight's difficult work. Phrases were heard about a "wheat deal in the sky," a "quarter billion dollar space handshake," or the bizarre posture of "our-arms-around-their-shoulders-and-their-hands-in-our-pockets."

Most well-informed observers agree that it wasn't anywhere near that bad. During ASTP no hardware changed hands. Procedural and managerial techniques were discussed, but each side already had its own way of doing space business. The institutionalized Soviet fetish for secrecy and compartmentalization probably made it impossible for the Russians to adopt any significant NASA techniques anyhow.

And by now it has been a decade since the Soviet engineers had this open window into the American manned space program. But there is no sign in their own current program of any "stolen secrets" or "purloined procedures." The French pilots who trained in Moscow for the guest cosmonaut mission in the summer of 1982 became quite familiar with Soviet space technology and later insisted it was all homegrown and not imported to any noticeable degree.

Both sides did lose something during ASTP—many of their misconceptions about each other's space programs and advanced technology. That was no loss. We could do well to undergo such a purge again, and as soon as possible.

A mutual familiarity with each other's manned space program today is bound to be salutary. The current whipped-up hysteria in the USSR about the "militarization of the American space program" and the perceived intentions for space-wars applications of the Space Shuttle may be partly a hypocritical smokescreen for Moscow's own space weapons programs (see Chapter 14). But in large part it could well stem from genuine anxieties and misperceptions. Grossly exaggerated Western press accounts of death rays, space bombs, and satellite kidnappings are bound to play on ingrained Russian paranoia and ideologically based Soviet fears. These frightening technologies are much further from reality than the Russians act like they believe, and an official joint space effort could deliver that message persuasively. Conversely, there has been parallel anxiety in the West about Soviet manned military reconnaissance activities in orbit and possibly some weapons-related tests as well. If such fears really are baseless on both sides, everybody benefits when untruths and misperceptions are corrected—a process which would be greatly strengthened during preparation for joint manned missions.

An American call for renewed manned space cooperation—starting with exchanging guest spacemen and progressing to physical linkups of space vehicles in orbit—would have bountiful political and diplomatic advantages. It would finesse the current Soviet propaganda offensive about a "renewed cold war psychosis." Shuttle-Salyut could graphically illustrate, at practically no cost, America's willingness to engage in substantive, mutually beneficial negotiations and cooperation despite political differences, where a fair and balanced deal could be assured.

The Soviets have given some hints that they would be receptive to such proposals. Writing in *Sovetskaya Rossiya* on April 22, 1983, Grigoriy Khozin (a specialist at the Institute for Study of America) recalled with nostalgia the cooperative spirit of the mid-1970s. "American scientists and the leaders of the U.S. space program almost unanimously advocated the further broadening of cooperation with the Soviet Union," he pointed out with some justification. "In September 1975 NASA Deputy Director George Low named among the tasks which the United States might begin resolving in the very near future the creation of a large orbital station. . . . At the same time he emphasized that the Soviet Union's participation in these projects would substantially improve their chance of success." Said Low, according to Khozin's approving recollection, "We must pool our efforts—I consider this to be the correct path." Khozin then approvingly referred to recent public statements by Senator Matsunaga and by ASTP Apollo commander Thomas Stafford praising such cooperation. "We must give the Staffords and Leonovs [the commander of the Soviet ship] of this planet one more chance."

In December 1983, Radio Moscow space commentator Boris Belitskiy responded postively to an alleged call for such a mission from a listener named "Edward Sandberg of Kenosha, Wisconsin." Said Belitskiy, "I think such a joint flight would be a fine thing for space exploration and it would be a fine thing for peace. . . . This joint project was unilaterally suspended by the American administration. As Valeriy Kubasov [one of the *Apollo-Soyuz* cosmonauts in 1975] said in July, everyone would gain if this joint project were resumed and most of all the general international situation would gain."

And in Washington, a number of politicians have begun to call for projects along these lines. Matsunaga's speech on March 10, 1983, was the most noteworthy: he called for a resolution to "renew the 1977 agreement on space cooperation" and "initiate talks for cooperative East-West ventures in space." Letters from numerous scientists who supported Matsunaga's initiative were appended to a bill. One letter was from Deke Slayton, a crewmember of the American side of *Apollo-Soyuz*. He wrote, "I am happy to have the opportunity to support your bill. . . . Coupling our demonstrated leadership in reusable launch and re-entry vehicles with the USSR's long duration large space station objectives [is an obvious possibility]." In early 1984, Matsunaga reintroduced the bill.

In July 1983, NASA Administrator James Beggs announced his general agreement. At a National Press Club luncheon for the STS-7 astronauts, in Washington, Beggs responded to questions about the possibility: "If they are willing to open their program a little bit more, then we would welcome a cooperative program and may be able to proceed. . . . Then we will cooperate again in a human activity. It may be something that would benefit us as much as it would benefit them. But more than that, it would probably benefit the world in a great way. So we would like to do it again—it's very difficult, but we'll try."

Symbiosis has been a time-tested and valuable trick of the trade for life on Earth, often even between mutually antagonistic organisms.

Perhaps that technique can move out into space. The 1980s are a time for learning to operate both independently and jointly in near-Earth orbits. Lessons learned could eventually be applied in the 1990s to a new deep-space exploratory mood. And if the Soviets and Americans and other space powers have learned through long practice how to productively cooperate in space, nothing—not even Mars—could be beyond their synergistic reach.

13

The Soviet Propaganda Blitz

Space cooperation must be founded on trust, and trust must be founded on truth. Perhaps the greatest roadblock to expanded U.S./USSR space cooperation is not an intransigent American government but a Soviet side immersed in misconception, distortion, and paranoia.

SOVIET PARANOIA

The cosmonaut's crewcut was tinged with gray, and his broad face was heavily lined. Georgiy Grechko, who had spent two long orbital duty tours aboard Salyut space stations, had an urgent message for his countrymen. In August 1983, he told a Soviet television audience, "We know that sights for laser weapons have already been tested on the first shuttle craft—and there are plans to deploy antisatellite systems in space." Grechko evinced great concern over these developments and struck an emotional chord among his viewing audience by appealing to the United States "as a cosmonaut and as a father" not to darken the skies with space weapons.

The speech was a genuine puzzle to Western observers. It was characteristic of dismaying new heights in the official Soviet smear campaign against the American space program, involving as it now did some of the USSR's most beloved and credible people: their space pilots and engineers.

But the biggest puzzle was that Grechko—and any other informed Soviet experts, of which there were probably very few—knew that the charges were counterfeit. It was Russia, not America, which had been testing orbital weapons such as "killer satellites." And the space shuttle, which together with its astronaut crews had become the target of an exceptionally vitriolic Soviet propaganda blitz, was innocent in the weapons-testing accusation.

SOVIET PROPAGANDA

The cosmic camaraderie between American astronauts and Soviet cosmonauts, which reached its apogee during the Apollo-Soyuz Test Project a decade ago, has seriously declined. Although good personal relations have in the past been fairly independent of international diplomatic conditions, this new situation has been the direct result of a Soviet decision to use cosmonauts as mouthpieces for Soviet propaganda to a much greater extent than in the past. This includes personal attacks on American astronauts, by name, and general assertions—exaggerated and often entirely false—about the alleged space militarization of the American manned space program. Omniously, the need for a justifiable Soviet "response" is also hinted at.

Such a policy was reflected in a statement attributed to General Vladimir Shatalov, commander of the Soviet space pilot corps. Shortly before the first space shuttle mission, Shatalov said: "We Soviet people, in particular cosmonauts, are pained to hear that some people in the United States are trying to use space technology for military purposes." In turn, Western experts were pained by the cosmic scale of Shatalov's hypocrisy when they considered the massive scope of Soviet military space activities and in particular of Soviet orbital weapons testing—an activity avoided entirely by the United States, except in the fantasies of Soviet propagandists.

During the Cold War in the early 1960s, nasty accusations against American astronauts were hardly surprising. One of the most active cosmonauts in this regard was Gherman Titov, the second man to orbit the Earth. Subsequently grounded for medical reasons, he seems to have vented his frustrations by attacking the character of American astronauts in general.

Titov had his own opinions about the "right stuff" possessed by

Anti-American space cartoons stress alleged "American space war threat" and militarization. (Illustrations reproduced from *Pravda* and *Izvestia*.)

his American counterparts. At one point, he told a press conference: "What motivations bring an American pilot to the launch pad? It is to gain fame, to make a fortune, to become a petit bourgeois exploiter all his own." Considering Titov's own status as a space has-been, that comment may have reflected more envy than enmity.

Shortly before the first launch of the *Columbia* in April 1981, the Soviet press began a cry of alarm over the background of copilot Robert Crippen. He was "a representative of a new profession—military astronaut," complained the TASS news agency. "The inclusion in the crew of a specialist of this type is added evidence of the big attention shown by the Pentagon," TASS continued. Unfortunately for the credibility of this propaganda blast, it seemed to have been based only on Crippen's background twelve years before as an astronaut in the cancelled USAF Manned Orbiting Laboratory (MOL) space station program. Since *Columbia*'s first mission, Crippen commanded STS-7 in June 1983 and two space shuttle missions in 1984—and none of them had any Pentagon connections.

Another space shuttle astronaut attacked by name in the Soviet press is Jack Lousma, commander of the third mission (and a veteran of the Skylab program in 1973). Lousma, since retired from space flying, was quoted as approving the development of the U.S. antisatellite missile system—which he had indeed said, in response to a journalist's question, with the understanding that the U.S. system had been first provoked by the Soviet killer-satellite program. The NOVOSTI news agency writer who reported this interview pretended to be horrified (while carefully excising any reference to the Soviet killer satellite, a program Moscow officially denies). "How can a space pilot say such an irresponsible thing?" the Russian demanded. "And he's a mission commander, too!" Obviously the Soviet audience was being warned to watch out for such "dangerous astronauts."

The most dangerous astronauts so far, if the Soviet-bloc media is to be believed, are Ken Mattingly and Hank Hartsfield of the DoD's STS-4 mission in mid-1982 (both men commanded new missions in 1984, and Mattingly's was again a DoD flight). During the seven days Mattingly and Hartsfield were in orbit, the Soviet press raised horrified alarms over the laser weapons which they were supposedly testing in space (they weren't). Their "warmongering cargo of destruction and death" was compared to "purely peaceful" Soviet space activities (except that Russia had tested a killer satellite only a few weeks before, and then lied about it).

In constrast, Soviet cosmonauts have been repeatedly trotted out to testify to their own love of peace—and of their opposition to nasty American space activities.

Cosmonaut Aleksey Leonov, speaking at the launching of a trio of Soviet space pilots in April 1983, pontificated: "The Soviet space program has always been aimed at resolving peaceful scientific and national economic tasks. I, who participated in the Soyuz-Apollo program, well remember meeting U.S. scientists and astronauts. During the meetings, the Soviet side repeatedly underscored the fact that space must never be allowed to be used for deploying weapons. This position meets the aspirations of all the peace-loving forces of the ᵖplanet. In sharp contrast to this background are the efforts by the current U.S. administration for the forced militarization of space." Later, in a congratulatory telegram to Sally Ride, Russia's first spacewoman, Valentina Tereshkova, announced her desire that space "should remain peaceful for all times, free of any kind of weaponry"—like the space shuttle, presumably. The cosmonauts clearly were omitting mention of many years of Soviet tests of orbital H-bombs and killer satellites—although it's hard to believe that they never had heard of them.

Other cosmonauts have obediently taken up the refrain. In the June 1983 issue of *Aviation and Cosmonautics*, three-time space veteran Col. Valeriy Bykovskiy wrote an emotional appeal. He denounced the "insane plans" of American militarists "who want to rule the universe," and continued: "Truly, war in space is much more dangerous than on Earth, since its consequences can lead via unavoidable processes to the end of life on our planet. Wherever is that common sense on which Americans pride themselves? Why does their president not think about the future of his nation, about the children?" Bykovskiy then denied that the Soviet space program had any military purposes. Indeed, according to cosmonaut Vladimir Dzhanibekov, the U.S. space shuttle program "responds to a great extent to the interests of war, not peace"—and any transfer of space technology for military purposes is "pernicious."

Particularly ironic was an appeal from cosmonaut Boris Volynov in August 1981: "The resolution of many global problems directly depends on successes in the exploration and large-scale use of space—but, I stress, only for peaceful purposes." In 1976, Volynov had been the commander of the *Salyut 5* space station, generally considered in the West to have been largely a military reconnaissance vehicle (even photographs and drawings of it are still military secrets in Moscow).

The claim that laser weapons had already been tested aboard space shuttle missions (both on STS-1 and STS-4) is a common theme; Grechko hardly made it up himself. It had first appeared in the January 9, 1981 edition of *Pravda*, where a TASS correspondent in Washington referred to an article in the *Baltimore Sun*. "One of the first tasks of the crew of the Shuttle, after placement in earth orbit in March of this year, will be the testing of the reliability of an aiming device for a laser weapon," wrote TASS. The project, according to the Russian correspondent, was code named Talon Gold.

The charge was repeated in other Soviet newspapers and over Radio Moscow, in both domestic and foreign broadcasts. News of the charge was reported by Western wire services from Moscow and subsequently re-echoed in the Soviet press as confirmation (with the detail that is was only an account of a "Soviet claim" carefully omitted).

And once the mission landed, on April 14, Soviet papers reported that the tests had indeed taken place. "The American astronauts tested a sight for laser guns on *Columbia*'s very first flight," thundered the April 24 issue of *Novoye Vremya*.

The fuss could have been avoided if anyone had really read the *Baltimore Sun*. The article in the January 7 edition was actually a Reuters dispatch from Washington, and it led off by asserting that "one of the space shuttle's early missions will be to test an aiming device for a space-based laser weapon." The dispatch quoted "congressional sources" in claiming that the Talon Gold test would occur early in the flights of the shuttle program—but not specifically on the first mission, as TASS misquoted the article.

The source of the Reuters account is harder to track down, because it is inaccurate. Talon Gold is indeed a real project, one of three associated with the Pentagon's space laser research effort. But sources in Washington, confirmed by Air Force spokesmen, make it clear that the project was many years from flying in 1981 and is still years in the future as of today. Dr. Robert Cooper, head of DARPA, indicated publicly in mid-1983 that a likely flight date was 1987–88.

The leading candidate for the "congressional source" for the Reuters mistake is the August 28, 1980 issue of the *Congressional Record*. In it, a congressman included the text of an article by David Ritchie in *Inquiry* magazine (September 1, 1980). Ritchie wrote: "DoD plans to put a huge laser weapon in orbit in the next few years. The project, code-named Talon Gold, is a scaled-down version of Darth Vader's Death Star. . . . DoD plans a laser test on one of the early space shuttle missions." This phraseology was strikingly similar to that later used by Reuters News Agency.

Ritchie, whose anti-DoD inclinations are plainly visible in the article, was obviously mistaken in his account of the schedule and the scope of the Talon Gold project. Since Reuters repeated the mistakes about the imminence of the laser weapon component testing, it is reasonable to suspect that Ritchie's errors were the ultimate source of all the fuss—all the way up to the cosmonauts pleading with America to "think of the children."

What seems to be happening is a deliberate preparation of Soviet public attitudes allowing the use of force against hitherto-respected American astronauts, all in the name of "self-defense." Whatever the USSR is about to be forced to do, so the argument goes, it will have been entirely justified by "American provocations." Any regrettable violence will only be a defensive preemption!

"Large-scale preparations are being made for a possible use of shuttle-type ships in combat," claimed TASS in March 1982. An official Soviet booklet, *Whence the Threat to Peace* (issued in 1982 in response to a Pentagon booklet on the Soviet military threat) wrote: "The Pentagon plans to use manned Shuttle spaceships as a space attack system. Various reconnaissance facilities and weaponry to hit space targets, including laser, are being developed for them."

Writing on the "Strategy of Space Madness" in the Soviet Army newspaper *Krasnaya Zvezda* (*Red Star*), Col. Eng. M. Rebrov referred to the "madman's delusions of the Pentagon maniacs"—"Military bases above the human planet, bunkers among the lunar craters, powerful laser guns aimed at Earth, satellites colliding in orbit, blinding flashes, lethal particle beams, the destruction of all living things. . . . This is an everyday matter in the Pengaton's militarist preparations." Rebrov continues: "In connection with the first flights by *Columbia* and *Challenger*, the Pentagon is already elaborating the most delirious and inhuman plans, a fever of mad ideas. . . . The task of preventing an arms race in space is becoming more and more acute, to halt the unbridled maniacs of space war."

Moscow's *International Affairs* magazine (November 1981) was explicit: "With the Space Shuttle, the U.S. Air Force can diversify the means of combat actions for striking targets in outer space, in the air, and on land. Equipped with nuclear weapons, they will act as a sort of outer space bomber."

The USSR armed forces journal *Sovietskiy Voin* (*Soviet Warrior*) told its readers: "The predatory looks that the American military clique is now turning to space are nothing but an attempt to use the greatest achievements of human intelligence to unleash new thermonuclear war for world supremacy. Using their new space weapons, the malignant people in the Pentagon intend to roll back the pages of history."

But the bland truth is that there are no lasers being tested, no bombs being carried, no kidnappings planned. Perhaps the Soviet fear is generated by what they know they would like to do if they had a spaceship like the *Columbia*. Or perhaps it is merely instinctive, unthinking xenophobia and irrationality, traits long cultivated officially.

This Soviet response can be considered menacing. "Pentagon militarists ought not to count on the USSR allowing them to turn outer space into a U.S. testing range for conducting preparations for war against socialist countries," intoned TASS military commentator Vladimir Bogachov in 1982. In *Izvestia*, analyst A. Krasikov wrote, "It is quite obvious that the USSR will not permit the United States to become the military master of space." In mid-1982, Brezhnev had warned, "The USSR will find the ability to rapidly and effectively meet any challenge hurled at us." Andropov made the same warning about space a year later: "All attempts to gain military superiority over the USSR are in vain. The Soviet Union will never allow this, it will never find itself unarmed in the face of any threat."

In light of these currents—the attacks on American astronauts as theatening the security of the USSR, and the promise to take action against such threats—perhaps another cosmonaut's speech needs more attention. At a ceremony in Ulaan Baatar in honor of Mongolia's first (and only) cosmonaut, veteran Russian cosmonaut General Georgiy Beregovoy noted that "only cosmonauts have the right to fly over the territory of another country without a passport"—but, he continued ominously, "this imposes certain responsibilities on them." Just what the USSR would do if it perceived overflying astronauts as being "irresponsible" is impossible to predict, but in the light of officially exacerbated Soviet public xenophobia over the KAL-007 atrocity, and the use of cosmonauts in a public relations campaign to smear astronauts, some anxiety may be called for. In any case, it's clear that cosmic camaraderie has hit a new low, propelled by deliberate Soviet attempts which may—or may not—have dangerous implications.

In Grechko's televised speech, he made the probably sincere appeal to American astronauts that "we will shake hands in space and not look at each other through gunsights, that we will not exchange laser blows but exchange information." Sadly, the cosmonaut's speech—and dozens of other public statements like it—are instead really setting the stage for just the kind of orbital confrontation the cosmonauts profess to want to avoid. If they really want to prevent such a confrontation, the cosmonauts could begin by stopping their cosmic deceptions and starting an exchange of authentic information—an extremely unlikely development.

14

24-Hour Astronaut Service

Nine times, manned spaceships in orbit have fired propulsion stages to push themselves much farther out from Earth. Between 1968 and 1972, these Apollo vehicles were headed for the Moon, the only visible target outbound from Earth.

The next time such a maneuver occurs, perhaps sometime in the 1990s, the astronauts may actually be aiming for an entirely different target, a special zone between Earth and the Moon. But whatever the practical justification for this alternate new target, the old target will remain in mind—and within range.

The astronaut's-eye view of Earth comes in two variants. In the more common, witnessed by more than one hundred and twenty men and women, the planet is seen from an altitude of several hundred miles. Its surface zooms by below, with landmarks staying in sight for no more than a minute or two before passing far astern. In the other view, experienced by only twenty-one men and not likely to be repeated much before the turn of the century, Earth is a full, round disk, coverable with a hand, seen from the porthole of a Moon-bound or Earth-returning Apollo spacecraft.

Both views create their own mental impacts. From low-Earth orbit, (or LEO), space crews remark on the wide expanses of Earth's surface and how unnoticeable are the works of humankind. From lunar trajectories, the astronauts were struck by the immensity of black space and the stark contrast with the green Earth.

GEOSTATIONARY ORBITS

There is another family of orbits in space, as yet unfrequented by human beings, which promises both new practical benefits and new views of Earth. These are the so-called geostationary orbits, or GEO. Located in a ring about 36,000 kilometers (22,000 miles) above Earth's equator, these are points in which satellites following circular orbits circle Earth at the same speed Earth rotates. That is, they would take twenty-four hours to circle the planet (in LEO it takes only ninety minutes)—the same time it takes a point on Earth's surface to be carried fully around once. Hence, to any point on Earth's surface, these kinds of satellites appear to hang motionless in the sky.

As is usual with fundamental concepts of space flight, the visionary Russian theorist Konstantin Tsiolkovskiy first noted the existence of such "24-hour orbits," in the 1920s. In 1947, Arthur C. Clarke proposed placing radio relay satellites there, and in 1961 the first such probe—called Syncom for *synchronous communication*—was launched. Today, more than a hundred active satellites and an equal number of derelicts are strung along the full "great circle" in a man-made ring around Earth's equator.

Besides communications relay satellites, surveillance satellites for both military and civil purposes are also operating in GEO. Infrared sensors, hanging continuously above the Indian Ocean, peer northward for a clear constant view of the entire Eurasian land mass. At other points, meteorological monitors send back full-disk imagery of an entire hemisphere's cloud cover.

The road to GEO has recently been shown to be a rocky one. Satellites such as NASA's Tracking and Data Relay Satellite (TDRS), along with payloads for Western Union and the Indonesian government, went astray in 1983–84 due to booster malfunctions while on their way from LEO to GEO.

A satellite jettisoned from the space shuttle is moving at about 25,000 feet per second, about 180 miles above Earth's surface. To boost itself into a transfer orbit reaching the target altitude, the satellite must gain an additional 8,100 fps from an attached rocket stage. Upon reaching the high point in the new elliptical orbit, the satellite would commence falling back toward Earth unless a second rocket burn of 5,900 fps were then performed. That burn circularizes the satellite's new orbit and also "turns the corner," or changes the plane of the orbit from that of the transfer orbit (usually equivalent to the latitude of the launch site) to a perfectly equatorial one.

THE GEO ENVIRONMENT

The conditions at these orbits are quite different from those experienced closer to Earth. "Space" may seem the same all over, but it's not—and spacecraft intended for operation at GEO, together with potential passengers, had better plan on accommodating themselves to the differences.

The first difference is the radiation environment. Above Earth's magnetosphere, satellites at GEO are subjected to the full force of solar and cosmic radiation and charged particles. These are not attenuated by the Earth's magnetic field; they bathe the space vehicles in unadulterated interplanetary radiations. The solar high-energy particles can induce electric charge buildup across the structures of the vehicles, leading to sparking within the vehicle's electronics (in the 1960s, many such satellites were lost when spurious computer commands were caused by such signals). If humans were to venture here, they would need "storm shelters" against solar flares, and they could not tarry long unprotected in the face of the continuous high level of normal radiation.

Sunlight is also stronger, not because the vehicles are closer to the Sun but because for months at a time the Sun never sets. Since Earth's equator is inclined twenty-three degrees to its orbital plane, a satellite circling high over the equator usually passes in front of and above/below Earth, or behind and below/above Earth's shadow cone. Only within a month of the spring and fall equinoxes does a GEO satellite dip into Earth's shadow—and then for periods of not longer than one hour out of twenty-four.

Nor are all points along the GEO necklace equal in stability and desirability. Due to lumpiness in Earth's outer mantle, and resulting gravitational variations, satellites at GEO tend to slowly drift toward "low" spots over specific longitudes. To avoid a cosmic collision, active satellites carry "station-keeping" propellant for small onboard rockets, which are fired periodically to keep the satellite in its assigned two-degree-wide slot.

So the environment at GEO is unique from that experienced nearer Earth. This usually causes design problems. But the advantages are worth it.

The view is new and spectacular: Earth fills an arc twenty degrees across in the sky (about the same size as a basketball held at arms length), and it runs through its phases from "new" to "full" to "new" again every twenty-four hours. When new, the unlit Earth is hardly visible in bright sunlight bathing the geosynchronous satellite. When full, it should be spectacular, casting shadows five thousand times brighter than full moon on Earth. During the occasional eclipses (occurring in groups six months apart), the whole Earth is surrounded by a luminescent ring of refracted light through the atmosphere. A passenger in such a satellite could continually watch weather patterns develop over hours and days, giving the impression of a living, breathing planet.

GEO EXPEDITION STUDIES

The spacecraft which will in the future operate in these locations will grow larger and more complex. Very wide antennas and large-scale optical devices require extremely fine adjustments and alignments. Already, there have been suggestions that future GEO platforms serving several dozen users will be precisely assembled by space shuttle crews and then gently pushed out toward their operating altitudes.

Once there, any future servicing (such as refuelling, or replacement of electronics boxes) could be done by robot "monkeys" sent up from LEO. But for such operations, the problems of remote control and of control lag (more than half a second for a round trip of a command in response to a stimulus) suggest that someday, under critical situations, people will have to venture out to GEO, too.

A NASA STUDY

NASA engineers in Houston recently completed a study on just how such a mission could be conducted.

The basic flight plan called for two astronauts to spend several days at GEO. They would launch from the payload bay of a space shuttle, coast upward for five hours, and then rendezvous with the target space platform. Several days later, they would fire onboard rockets to fall back toward Earth where they would skim through the upper atmosphere to kill off excess speed and change plane, as required, to swoop up alongside the still-orbiting shuttle. After more maneuvers to complete a rendezvous, the crew would transfer to the shuttle for return to Earth. The GEO ferry capsule would remain in orbit for future reuse.

The engineers found that the mission could be accomplished with a five-ton Apollo-class command module atop an uprated Centaur rocket stage. The command module would have to be already in orbit or brought up on a different flight; one entire shuttle mission would have to be dedicated to carrying the fully fueled Centaur alone.

The old Apollo landed in mid-ocean, cushioned by parachutes and water. But the seawater ruined any chance of reusability (the original Apollo design called for three reflights of the hardware, but this was abandoned when splashdown mode was selected). If the new proposed capsule is stripped of its parachutes and flotation gear and is loaded with extra maneuvering fuel, a precision guidance system can steer it into a retrievable parking orbit instead of a surface-impact trajectory. This was the route chosen by the NASA engineers: "Aerobraking eliminates the development of water landing systems and water recovery operations," their report concluded. Even with the extra maneuvering, it made for a lighter command module.

Estimates were that such a mission could be flown by 1990, at a cost of about a billion and a half dollars. Subsequent flights would each cost about one-tenth this much.

A companion vehicle to the GEO sortie tug would be the "line shack." This twelve-ton module would remain near GEO, and it would contain spartan living quarters for visiting teams of astronauts. Since most standard tools and the heavy airlock could be left attached to the line shack, the weight of the sortie ferry capsule could be reduced by about one-third. But the line shack would cost another billion and a half dollars to build and to launch.

In late 1983 the study results were presented to NASA headquarters in Washington. Reportedly, the opinion there was that the study was nice, thank you very much, but it was too dangerous and too expensive in terms of any conceivable mission in the next ten years.

An Alternative Plan

Other space engineers have investigated the same type of mission which NASA analyzed. At Eagle Engineering in Houston, Hu Davis has long been an advocate of the development of manned geosynchronous capability. But he sees little purpose in merely making a sortie out there to show it can be done; his plans involve developing a significant in-place work capacity.

"I want two things beyond a bare-bones demonstration," Davis told me early in 1984.

The first deals with mission capability, particularly endurance and maneuverability: "We need to be able to cover a significant fraction of the geostationary arc while we're there," he asserted. "And we need to stay for a week or more." Since no space shuttle could tarry in orbit that long, the GEO ferry capsule would have to contain its own earth landing system. This would probably be a parachute for a dry land thumpdown.

Second, "I want at least one ton of discretionary payload," demands Davis. No stripped-down capsule could carry enough cargo to do anything useful once it got to GEO, he claims, but with a cargo capability, sortie vehicles would be able to bring up hydrazine propellant, new batteries, spare apparatus, and improved radio transponders for installation on degrading communications relay satellites. "These are useful payloads to enhance the value of orbital assets," continued Davis.

The larger vehicle would need more launch power. For each GEO expedition, Davis foresees one full shuttle launch and one full launch of a Shuttle-Derived Vehicle, a heavy unmanned space cargo carrier capable of putting double-sized payloads into orbit (see Chapter 5).

Such a capability would allow some really useful missions, for flights at least once a year, Davis believes. "Most GEO satellites that die, starve," he pointed out. Refuelling with new propellants could restore such satellites to life.

But currently, most users of GEO satellites figure that after seven to ten years of lifetime, the satellite's technology is so obsolete that it makes more economic sense to replace it entirely. "Sure there's no incentive now," Davis pointed out, "because the communications satellite manufacturing companies sell new satellites and replacement satellites."

Instead, Davis suggests that the older satellites could be sold to customers with less need for the very latest technologies. Or if high technology were the object, the satellites could have their radio transponders swapped out while leaving the still-serviceable power, attitude, command, and control systems.

And sooner or later the large GEO platforms now being planned will demand this servicing capability, first by robot "teleoperators" but eventually by on-site humans as well. Davis sees this coming by the end of the 1990s, and that is when he expects to see astronauts visiting GEO space. "It's an implicit part of NASA's new space station program," he claims, since the placing of the permanent station at a twenty-eight degree inclination is convenient for voyaging deeper into space, while sacrificing coverage of Earth surface targets farther north or south than twenty-eight degrees. "Washington has called for a bolder,

Manned spacecraft departs low-Earth orbit for sortie twenty thousand miles higher, to repair geosynchronous communications relay satellites. A slightly different flight plan could take the capsule all the way to lunar orbit at no added cost in fuel. (Photography courtesy of NASA.)

more innovative, and broader reaching space program,'' Davis recalls, ''and this could be part of it.''

Davis's eyes twinkled. ''And besides,'' he admitted, ''GEO sortie is the backdoor to the Moon.''

The cold numbers of space navigation support his pleasantly surprising claim. In terms of spacecraft ''delta-V'' (velocity change), which relates directly to fuel load, it is actually cheaper to go all the way to the Moon, orbit a hundred miles above it, and return to Earth, than it is to reach and return from GEO. It takes longer, three days each way instead of six hours, but any GEO sortie ferry which is built to support astronauts for a week at GEO could just as easily support them on a return voyage to lunar orbit.

So whichever capability is ultimately commissioned first, the other will follow at little actual cost. If commercial reasons require manned operations at GEO, they will accidentally reopen the road to the Moon.

But for the space engineers who have been planning and justifying such a new mission, with carefully constructed rationales for profit and advantage of GEO sorties, it is no accident. They can add the delta-V numbers. And they know what the real goal is, and how close it actually may be.

15

Flights to Other Worlds

THE MOON

The last man on the Moon, more than one dozen years ago, promised to return. "I take man's last steps from the surface for some time to come—but we believe not too long into the future," mused astronaut Gene Cernan out loud as he paused before climbing his spacecraft's ladder. "We leave the Moon as we came, and God willing, as we shall return, with peace and hope for all mankind. Godspeed from the crew of *Apollo 17!*"

The thought of sending human beings back to the Moon has receded farther and farther since that moment, with continuing shortages of funds, imagination, and boldness. Ironically, earthmen are now probably farther from landing on the Moon now than they were when the first *Sputnik* was launched more than a quarter century ago. Then, it took less than twelve years to accomplish the first landing mission. Now, the idea of astronauts returning to the Moon within twelve years in the future, by 1996–97, is generally considered absurd and unrealistic.

The Antarctic Model

However, all is not lost. Many historians of exploration, who tend to take the long view of things, consider that lunar exploration may follow the "Antarctic model." Under this view, the Apollo landings were the equivalent of the Amundsen/Scott South Pole races of the early 1900s. They were followed by aircraft overflights a few decades later, and ultimately, in 1947, by Operation Deep Freeze, the beginnings of a massively supported effort to establish permanent scientific stations on the coast of Antarctica and, finally in 1957, at the South Pole itself.

The Moon may undergo a similar progression of exploratory phases. New unmanned orbital missions, completing mapping and geochemical surveys, may occur within ten years. They could be followed, near the turn of the century, with a major project to set up one or more permanent scientific stations. By then, the technology of the space shuttle and Shuttle-Derived Vehicles, together with upper stages developed and tested for use in near-Earth space (the Orbital Transfer Vehicles, or OTVs), will make the expeditions feasible and cheap—a fraction of the original cost of Apollo, for a tremendous expansion of capability.

The scientific and applications foundations for just such an eventuality are even now being laid by half a dozen separate and so far uncoordinated groups of specialists and enthusiasts, in NASA, the university community, the aerospace industry, and among private groups. In March 1983 in Houston, while the fourteenth Lunar and Planetary Science Conference convened to analyze recent results from interplanetary studies, initial organizing meetings were held privately.

The Lunar Initiative

What was called the Lunar Initiative was unveiled in mid-1982 by a team of space scientists at the NASA Johnson Space Center in Houston. In full awareness of the moribund state of planetary science, the scientists recommended that the goal of a permanent lunar base be considered for sometime in the first decade of the next century. To develop the information base needed by 1990 to make a proper decision on placement, architecture, and a mission of such a facility, the scientists further suggested that a series of unmanned lunar missions be prepared. The first step, which should occur as soon as possible, would put a half-ton survey satellite into lunar orbit; it would be followed by remote-controlled rovers on the lunar surface.

Lunar orbit station could be an observatory and fuel depot to support large landing expeditions. (Photograph courtesy of NASA.)

Moonbase Benefits

Support for the moonbase idea also has appeared from a number of unexpected places. In May 1982, two Los Alamos National Laboratory physicists issued their own independent plan for such a base. According to Dr. Paul Keaton and Dr. Eric Gelfand, a large 24-person facility could be built by the end of the century for a program life cost of considerably less than was spent on the Apollo program in the 1960s. They called for "a national commitment for an International Research Laboratory on the Moon," and added that "a vigorous civilian program like that proposed here is our best guarantee that outer space will be used to strengthen our economy and address basic problems on Earth." This advocacy by a group of traditionally nonspace scientists took on added significance when it was realized that the White House Science Advisor is a former Los Alamos official (reportedly, he personally gave a "positive response" to the proposal—"so long as no money was involved"). Other nuclear scientists around the nation, particularly Dr. Edward Teller, subsequently voiced their own support for such a plan.

With the Moon as a research base, the space scientists foresaw major benefits both to astronomy (particularly, but not exclusively, radio astronomy and solar physics) and planetary geology (through a careful investigation of the Moon itself). But more far-reaching (and at the same time more immediate) uses were considered possible in the use of the Moon for a source of natural resources and as a role-player in national security.

"The Moon is still the keystone for understanding the terrestrial planets, including the Earth," noted Dr. Wendell W. Mendell of NASA. And despite the closeness and relative ease of access of Earth's natural satellite, ironically "Mars has been mapped more completely than the Moon," Mendell added.

"Occupation of the Moon is an integral part of manned activity in near-Earth space," the Houston report had stressed. Added Dr. Mendell, "A lunar base is the inevitable culmination of that activity. From simple considerations of booster performance it is evident that the development of the capability for manned activity in geosynchronous [24-hour] orbit also permits activity in near-Moon space."

One key resource which the Moon might provide is water. Theoretical considerations indicate that because the Moon's axis of rotation is nearly perpendidular to its path around the Sun, there should be deep craters in the polar regions inside of which sunlight has not fallen for billions of years. Water in the form of "dirty ice" should have gathered there from passing comets or from occasional outbursts from the Moon's deep interior.

It is to find such possible sources of water that scientists have long been urging the launching of an unmanned lunar polar orbiting probe. "Its scientific value is not questioned and its relatively low cost is well-known," Mendell pointed out. The probe would require about one quarter of a space shuttle mission (that is, it could share the cargo bay with a number of other payloads) and could be launched as early as 1989 if NASA convinces the government to fund it.

One drawback may have been the name of the probe. First it was LPO, reminiscent of a brand of dog food. Now the scientists want to call it LMS, for Lunar Mapping Satellite. But considering its mission and the wealth it is supposed to find, a zingy name such as Prospector might serve it better at budget hearings' time.

If water is found on the Moon, the cargo weight needed to support teams of astronauts on the surface would be dramatically reduced. The presence of water would also allow the on-site manufacture of rocket propellant (cryogenic hydrogen and oxygen), which would further reduce the operational costs by a factor of two or more.

Turn-of-the-century "lunar module" lands equipment for mining the surface for oxygen to be used (and sold) as rocket propellant. (Photograph courtesy of NASA.)

Perhaps in keeping with the perceived priorities of the current administration, some scientists have also pointed out the national security aspects of a moonbase. "We see the Moon as the ultimate high ground to protect near-Earth space for the benefit of the nation," noted Dr. Jeffrey Warner in Houston. Missions could include observation and communications, with the advantage over near-Earth satellites lying in the possibility of heavily shielding the lunar surface facilities and of attaching them firmly to bedrock foundations. This would solve two of the major current problems in near-Earth military applications satellites: those of survivability and of attitude and pointing control.

The Moon makes much less sense as a weapons platform (except perhaps in a space-to-space defensive application) because of its great distance from Earth. Nonetheless, Soviet propagandists have accused American military researchers at the Redstone Arsenal in Huntsville, Alabama, and at Strategic Air Command (SAC) headquarters at Offutt Field, Nebraska, of plotting to emplace laser weapons on the Moon in violation of international treaty. Spokesmen for both agencies denied the Soviet charges.

Transport to the Moon

The transportation problem to and from the Moon is not a particularly difficult one. In Apollo days, a 240,000-pound spacecraft was placed in low-parking orbit above Earth, where a propulsion stage pushed the vehicle out towards the Moon. Another propulsion firing put the whole craft in lunar orbit, from which a landing craft (the twenty-ton Lunar Module) descended to the surface. After a brief period of hurried exploration, the crew blasted off again to rejoin the waiting return spacecraft in lunar orbit; three days later, their capsule was floating in the Pacific Ocean.

Manned voyages to the Moon two decades from now will almost certainly make use of space shuttle–type technology, which by that time will include a "heavy lift vehicle" built of shuttle engines and tanks (and capable of hauling up to 180,000 pounds into orbit) and a high-energy upper stage (possibly based on an extension of cryogenic Centaur rocket technology). Also, the technique of *aerobraking* will probably have been perfected solely to support the return of manned and unmanned spacecraft from geosynchronous orbit. The technique calls for the use of an inflated balloon to act as a "drag brake" in the Earth's upper atmosphere, which helps the returning space vehicle slow down enough to enter a stable orbit—but not slow down too much and fall out of orbit and deep into the atmosphere. This allows the vehicle to dispense with a heavy heat shield since it can be safely picked up in parking orbit by a space shuttle mission.

To get to the Moon, several orbital transfer stages could be assembled in parking orbit, either on several shuttle missions or on one launch of the unmanned heavy vehicle. Another shuttle launch would carry the lunar landing stage itself (probably a modified transfer vehicle) and the crew cabin. Two stages, attached in parallel with the core vehicle, would push it out toward the Moon and then separate to return to parking orbit for reuse. The third unit would park itself in a stable orbit above the Moon while the landing vehicle made the descent, surface stay, and ascent. The entire craft would then return to Earth, using aerobraking to reach a safe parking orbit for eventual pickup.

The Staging Area

The actual operations connected with the resumption of manned lunar flight may depend heavily on the establishment of a permanent "staging area" near the Moon. The station would be a scientific platform in its own right, as well as a refuelling depot. Most American plans talk about placing the station in orbit a few hundred miles above the lunar surface. Soviet studies, however, suggest that the best location for a manned near-Moon station (and the Soviets could be serious about such a mission by the end of this decade) would be "L1," the first *Lagrange Point* which lies directly between Earth and the Moon. Although it is a theoretically unstable point, a recent Russian report claimed that a station could maintain position there for a very modest propellant expenditure.

An Economical Project

Wherever this station is placed, it will contribute to the economy and ease of future lunar exploration. The reusability of the stages, plus the fact that their original development expenses would be amortized over different missions, would both contribute to the significant economy of such a program. Apollo cost so very much because an entire space transportation capability had to be built from scratch. Returning to the Moon will be relatively inexpensive because most of the required capabilities will have already been bought and paid for!

"We won't go back to the Moon until it's easy to go," noted one top space official as the Apollo program drew to a close. Twenty years from now, it will have become easy—due to work being done now for other projects, with other justifications, and billed to other customers. Political decisions in the next ten years will determine whether the Apollo program constitutes a brief interlude in the long, dead history of the Moon, or instead a prelude to the establishment of a permanent human presence there at the turn of the century. And once people reoccupy the Moon, it will not be the end but the beginning of distant voyaging.

MARS

Mars glistens in the human imagination as brilliantly as it does in the night sky. Dreams of human visits to the red planet, however, faded in the 1970s along with the dying batteries of the Viking landers. Since then, the public mythos of flying to Mars has shifted significantly; where once the most popular books and movies on the subject described heroic expeditions into the unknown, the most notable Mars movie of the 1970s—*Capricorn One*—depicted a sordid cosmic hoax and the ultimate failure of space technology.

Surprisingly, however, things are looking up. Without fanfare, scientists have been assembling increasingly powerful reasons for manned expeditions to Mars while space engineers in the United States and the USSR have been developing the kind of flight hardware which could readily be diverted to the support of such an assignment. Optimists now believe that ambitious interplanetary missions could be mounted at the turn of the century with half the equivalent expenditure of the Apollo project.

Soviet space officials make no secret of their desire to make such flights. Late in 1979, cosmonauts Georgiy Beregovoy and Valeriy Ryumin told Soviet radio listeners that manned missions could begin in ten or fifteen years. A year later, after returning from his second six-month space flight in a row, the 41-year-old Ryumin announced his readiness to go up yet again: "If an expedition to Mars were being prepared and it should be necessary to hold a year-long stay in space as an intermediate step, I think that we would readily agree to such work." Chief Soviet space doctor Oleg Gazenko told European scientists the same thing in November 1980: "It is difficult to give an exact date for a flight to Mars. But I think the basic prerequisites for such a flight exist now. . . . Whether the flight happens in 10, 15, or 20 years, I cannot say. But I believe it will be before the year 2000."

Compared to landings on the nearby Moon, a Mars flight seems many times more difficult. The spaceship must operate for years instead of weeks; higher launching and landing speeds are involved; twice as

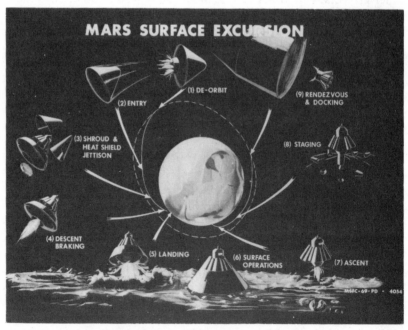

Mars landing sequence. (Photograph courtesy of NASA.)

Spacecraft to land people on Mars have already been conceptually designed. (Photograph courtesy of NASA.)

many crewmembers must be accommodated. The result is that a Mars-bound ship, departing from a parking orbit near the Earth, must weigh at least six times as much as a Moon-bound Apollo-Saturn. And, for safety's sake, most experts call for the launching of *two* six-man vehicles in parallel flight.

But there are some positive sides to the new deep-space target world. First, since Mars has a thin but significant atmosphere which can be used for aerodynamic braking, the landing itself can be relatively cheap. Secondly, Mars appears to have immense quantities of water ice which can be used by the visiting earthmen both for drinking and—after processing—for breathable oxygen. Lastly, Mars has two orbiting fuel depots (currently known as the moonlets Phobos and Deimos) available, after processing, for restocking rocket propellant needed for the homeward leg.

The Cost

Probably one of the most startling aspects of a hypothetical manned Mars expedition in the 1990s is its projected cost. While post-Apollo critics were bandying estimates of a hundred billion dollars or more (four to five times the cost of Apollo), actual engineering studies have produced far lower figures. These range from a half to a whole Apollo cost equivalent—and these kinds of studies have been reasonably accurate in the past.

Dr. Charles Sheldon, a leading space researcher at the Library of Congress, outlined the logic of this surprisingly inexpensive forecast in a special space report prepared in early 1978. "When the United States decided to send men to the Moon," Sheldon wrote, "it had to construct an entirely new capability. . . . Yet by the mid-1990s, the U.S. will have developed—purely on their own merits—three of the four types of new spacecraft needed to support a man-on-Mars mission: the Space Shuttle, a space station core module, and perhaps an efficient 'orbit-to-orbit tug.' " Only the fourth spacecraft, a deep-space injection stage, would have to be developed from scratch—it could be a nuclear engine, or an advanced propulsion system such as ion drive or an *electromagnetic catapult* (a *mass driver*), or just clustered Centaur engines burning liquid hydrogen. "Under these conditions," wrote Sheldon, "planetary flight falls drastically in cost, and even allowing for all the special payloads that might be required, a Mars expedition . . . might be priced at a figure far closer to $10 billion (1971 dollars) than to the former estimate of $100 billion."

An Obvious Target

But Carter administration policymakers didn't seem to be paying attention to their own government experts. "Nobody in Congress or the federal government or the public has put forward a case for a U.S. manned Mars mission," argued White House Science Advisor Dr. Frank Press in late 1978. "And if the Soviets decide to spend seventy billion to land men on Mars in five years, we say 'God bless them.' "

Those frequent Soviet statements about manned interplanetary flight as an eventual development in their space program are generally considered to be idle boasting from officials heady over the impressive series of recent space successes. The Soviet space station *Salyut 6* remained in orbital flight for more than five years, long enough to have journeyed to Mars and back three times had its launch vehicle packed more horsepower and its storage lockers more food and spare parts. The record-breaking cosmonaut endurance runs of the past six years have all but eliminated concern about the long-prophesied medical barriers to manned interplanetary flight. Meanwhile, the closed-cycle regenerative life-support systems being tested in space by cosmonauts are just the kind of equipment needed for installation on a Mars vehicle.

So even if the Soviet statements are just grandstanding, the Russians are now doing precisely the kinds of in-space tests which would be mandatory if they had already secretly commited themselves to such plans.

Spokesmen within NASA have also discretely voiced sentiments about the desirability of such an ultimate goal. Speaking to employees at the Marshall Space Center in Huntsville, Alabama, in March 1980, the then NASA administrator, Dr. Robert Frosch, was bullish on the concept: "My guess would be that we should be working on a new manned mission in the 1990s and Mars is the obvious target," he told the space team which had built the rockets that took men to the Moon. "It will be within our lifetime," he concluded.

In Houston, meanwhile, scientist-astronaut Dr. Karl Henize had just issued his own call for a manned Mars expedition on scientific grounds. "The ingenuity of life outstrips man's imagination," he wrote in March 1979. "Thus, life on Mars may indefinitely elude the limited tests and sensors that can be accommodated on unmanned probes. When the scientific objectives of a mission are complex and not easily pre-programmed, direct human exploration may, indeed, be the most cost-effective approach."

Such a Martian expedition may be primarily seeking scientific knowledge: Martian biochemistry, geology, mineralogy, meteorology and climatology—all of which could by comparison with Earth give scientists new insights into our home planet and its place in the Solar System. But other observers suggest other motivation for the flight: Dr. J. Peter Vajk, author of the pro-space *Doomsday Has Been Cancelled,* has pointed out that expanded near-Earth space industrialization activities in the 1990s would benefit tremendously from mining the easily accessible material resources of Phobos and Deimos and transporting the products back to space colonies orbiting the Earth-Moon system. A Mars sortie of this type could actually pay its own way, claimed Vajk.

A Cooperative Venture

A man-to-Mars expedition may even harvest more than mere scientific knowledge and valuable raw materials—it may find peace on Earth. Seriously, such a project may encourage—it may even have to rely on—increased international space cooperation and all the concomitant political compromise which would follow.

Due to the parallel complementary manned space programs of the United States and USSR in the 1980s, the world's two major spacefaring nations will each be able in the 1990s to produce unique components of a complex man-to-Mars transportation system, components assigned to both countries based on their individual specializations and flight experience. This would essentially split the mission cost; further reductions would come from the participation of additional countries contributing special pieces of equipment.

The crew habitation module, designed to keep the Mars-bound men and women alive over the long months in deep space, could be based on the Soviet Salyut space station. The Mars landing module would have to be a combination of Apollo Lunar Module and Viking Lander technology, making the United States the obvious choice to build it. Launching the vehicles piecemeal into low-Earth orbit could be accomplished by a mix of Soviet superboosters and American space shuttles, while the on-orbit assembly would be done at a Soviet space station where similar work had already been conducted to support Soviet manned lunar activity of the 1990s. The departure from low-Earth orbit could be accomplished via the use of nuclear-powered space tugs, reportedly under development by the Soviets (alternately, the United States might have a more advanced propulsion system in development by that time, too). And the Earth landing module would look a lot like an Apollo command module—so the United States should build that unit, too.

The result would be a truly international space venture which would last at least two years, tying Soviet and American attention together in one arena—and possibly spilling over into other areas as well.

A Long Mission

So much time is needed to get to Mars and back that it makes sense to stay there awhile. Early NASA studies of an "austere" mission envisioned at least two months on the surface. Some flight plans followed trajectories which required a layover of almost a year. Recent studies have suggested that as early as the second expedition, the astronauts should set up a base camp and wait for the arrival of a relief party 2½ years later. The costs of such a semipermanent settlement, according to recent analyses at several universities, need not be significantly higher than those of brief visits—if the Martian explorers can learn to "live off the land."

One specialist in this type of preliminary Mars colonization is physicist Penelope J. Boston, of the Laboratory for Atmospheric and Space Physics at the University of Colorado in Boulder. In a series of studies beginning in 1978, Boston and her colleagues have tried to define the processes by which Mars explorers could obtain water, air, and food from native Martian resources. Water appears to be easiest, and oxygen can be obtained from chemical reactions fed by plentiful carbon dioxide gas. Food is a more complex problem.

Crops could be grown in "Martian greenhouses," which are simply plastic structures inflated to several times the ordinary Martian surface pressure; laboratory experiments by Boston's group have found that many terrestrial life forms can be adapted to such conditions and could form the basis of an agricultural system rooted in Martian soil (one early promising plant was the radish—so the researchers humorously put out a call for any and all radish recipes they could get their hands on).

Meanwhile, other team specialists have examined more far-reaching and mind-boggling possibilities for Mars. According to graduate students Christopher McKay and Steven Welch, human settlers on Mars may eventually be able to alter the climate of the entire planet to make it more hospitable for human colonization. But such plans are centuries from realization. Martian greenhouses, however, could be feeding Mars's first human settlers within two decades (they could even be feeding some members of the University of Colorado study team since most of them are young graduate students with three or four decades of professional life ahead of them).

The most vocal current advocate of a manned Mars expedition is former Senator Jack Schmitt, acting director of the Division of Space Biomedicine at the Johnson Space Center. He has already published proposed space exploration schedules—his so-called Chronicles Plan—which call for Mars flights by the year 2000 (the same milestone tagged by Dr. Oleg Gazenko in Moscow). Speaking to space workers at the NASA center in Houston in December 1980, Schmitt repeated his call for deep-space American manned activity. "It might be helpful to realize," he pointed out, "that very probably the parents of the first native-born Martians are alive today."

The Mars Underground

Because of—or despite—the logic of such a project, this new space goal seems about to spring into the public's awareness again after a decade's hibernation. The idea of sending people to Mars, to visit or even to live there, has always had a powerful grip on the popular imagination; the television adaptation of Ray Bradbury's *Martian Chronicles* appealed to that imagination. The idea has been gathering surprising strength lately. A "Mars underground" of enthusiastic space professionals has already held two symposiums in Colorado.

Such interplanetary expeditions could set sail before the turn of the millenium. Although it will be the engineers—American, Soviet, and others—who will really "bend the metal" to build the ships, the mission itself has already taken form in the minds of dreamers and prophets, such as Gazenko, Schmitt, Boston, and Bradbury, and their predecessors and successors. The necessary political decisions are the very last hurdle between the dream and the reality—and arguments for a "go" are getting more compelling every year. Mars is waiting, glistening brighter.

16

Exploring the Asteroid Belt

The riches of the asteroids have been exploited by earthmen since prehistoric times when cavemen discovered metallic meteorites—chips off of asteroids—and fashioned from them their first tools and weapons. In the near future, widely venturing space prospectors will seek out the "mother lodes" far from Earth and extract the wealth which has been waiting beyond human reach for millions of years.

The other side of the asteroid coin is equally dramatic and important: avoidance of an excess of such asteroid material coming to earth unexpectedly. Giant meteor craters on all solid surfaced worlds of the Solar System are mute but eloquent testimony that we on Earth inhabit a "celestial falling rock zone," and it is only a matter of time before another "dis-asteroid" strikes. Rough calculations of the immense cost of such a cataclysm, multiplied by its small annual probability, give a figure of several billion dollars a year as a "reasonable" insurance expense to invest in discovering ways to prevent such an eventuality.

These then are "the carrot and the stick" to the question of asteroid exploration and exploitation. On one hand, asteroid resources can support expanded space activities and make far sighted investors (individuals, corporations, and nations) extremely rich; on the other hand, a failure to come to terms with the nature of asteroids will undoubtedly (but unpredictably) lead to vast devastation and death on Earth.

Both issues share a common technology and a common opening strategy. First, the asteroids nearest Earth must be discovered and charted—and today, despite a few dedicated efforts, very few of them have been. Second, the physical nature of these mysterious bodies must be understood. Third, advanced space transportation vehicles must be developed to bring both robots and people to some of the nearer asteroids, and then back again. Mining technology must be developed to process the raw materials into the desired products. And last, technologies must be developed to divert portions of these asteroids—or their entire bodies—onto new courses, either to bring them back close to Earth for exploitation or to nudge them safely off course when such a target is *not* desired.

Most of the asteroids—which range in size from a few hundred feet up to hundreds of miles across—follow stately and harmless orbits between Mars and Jupiter. They are evidently leftover scraps of planetary debris from the birth of the Solar System (most astronomers have finally abandoned the theory that they were fragments of a large exploded planet).

APOLLO OBJECTS

Those asteroids of most immediate interest, from the point of view of greed and fear, are on different paths. They come in much closer to the Sun, crossing the orbits of Mars, Earth, and on occasion even Venus and Mercury. Although some are a few miles across, others are only 300 feet across. And they probably number in the hundreds of thousands. The ones which cross Earth's orbit are called *Apollo objects*, named not in memory of the Moon missions but after the classical name given to the first one of its kind, discovered decades ago.

These Apollo objects are among the most accessible extraterrestrial bodies in the universe. The velocities needed to reach some of them are in some cases lower than that needed to get to the Moon's surface and back. Round-trip travel times, however, are significantly different: a week for the Moon, and three years for an asteroid. But for robots or for astronauts with advanced recycled life-support systems, that may not be a problem. The ease of access has led many spaceflight theorists to suggest that astronauts will visit asteroids long before they land on Mars, Ganymede, Mercury, or other larger and better-known worlds in the Solar System.

MINING THE ASTEROIDS

What will the first visitors to asteroids be looking for? It is easy enough to say wealth and resources, but first one needs a better understanding of what is really valuable in space, as opposed to on Earth. Only when the goal is well defined, and the tools have been built with that goal in mind, is success feasible.

Gold, silver, platinum, diamonds, and other precious materials have been sought by explorers on Earth for thousands of years. But spaceflight induces its own unique form of coinage.

The most valuable raw material currently being sent into space is oxygen—and to burn, not to breathe. More than one-third of the weight brought into orbit by moon-bound Apollo-Saturn rockets was cryogenic oxygen propellant. So expensive is the cost of transportation into orbit that the comparative difference in value of gold, say, and water, quickly disappears. And in fact many specialists would say that water is more important than gold since it can be converted into rocket fuel to get you home while gold will only weigh (or "mass") you down.

There have been studies which suggest that a 300-foot asteroid, weighing one million tons, would contain between three and ten billion dollars worth of gallium, germanium, platinum, and other precious metals. Under certain optimistic circumstances, it might become economically feasible to import such substances to Earth for commercial sale at a profit. And doubtlessly this will someday be done—but not at first.

The earliest asteroid mining expeditions will almost certainly be sent out to obtain materials most valuable for use in space itself. There, any space-derived resource has a natural cost advantage over material hauled painfully up the first few hundred miles of Earth's precipitous "gravity well." The most attractive resource, which would be offered for sale to orbiting space vehicles with empty fuel tanks, is the aforementioned substance, oxygen, which is for all intents and purposes free on Earth, In space, however, it will be the coin of the realm—rocket-style.

The extraction of the oxygen can occur in several stages. If there is water chemically locked into the asteroid's rocks, the material can be pulverized and then heated in solar ovens to steam off the water vapor. The water, in turn, can be electrolyzed using solar power to produce hydrogen and oxygen. Alternately, specific minerals can be

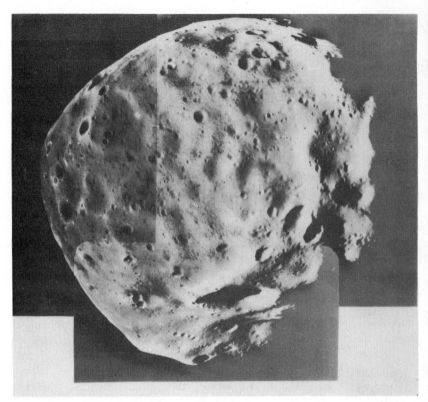

Typical asteroids probably look like this view of Phobos, inner moon of Mars. (Photograph courtesy of NASA.)

separated either magnetically or through centrifuging, and these minerals can in turn be "baked" to free their trapped water. In parallel, other processes are concentrating metallic ores for chemical refining. All processes share the need for machinery to bite off, pulverize, contain, and transport large hunks of the asteroid—and all under conditions of essentially zero gravity, full vacuum, and unpredictable sunlight.

Moving the asteroid back toward Earth can be done by throwing away parts of it. If rocket propellants have been manufactured, then a fraction of the cargo will have to be used to get the rest home. Nuclear blasts could shove the bulk of an asteroid in one direction by thrusting charred debris in the other direction. More gently, electromagnetic catapults (powered by sunlight) could eject a stream of waste material to coax the rest of the asteroid into an opposite direction—and estimates are that such a scheme could return about a third of the asteroid's original mass by feeding two-thirds of the mass into the catapult over a two- or three-year period.

Any of these techniques would serve equally well to deflect or disintegrate a small asteroid discovered to be on collision course with Earth (we know for certain that there are such asteroids, but so far we just don't know how big they are and where and when they will hit). Once again, numerous different applications in space share many of the same technologies.

In coming centuries, as in the past, both wealth and death will be descending on Earth from the asteroids. For the wealth, our own choice is limited to one of striving to share in that wealth or let other more far sighted societies harvest it to their benefit. For the death, our choice is limited to one of burying our heads in the sands of Earth since some will certainly die. The alternative is acting with foresight and thus reaping both life and fortune from beyond the sky.

Already, however, some asteroids have provided practical demonstrations of theoretical mathematics back on Earth.

LAGRANGIAN POINTS

No sooner had the eighteenth-century mathematician Lagrange demonstrated that there were five theoretically stable points in a two-body system, than other scientists began to wonder if such regions existed in reality. A century later, it was learned that they do—and that many are already occupied. In the following century, plans appeared to occupy others with man-made objects.

Trojan Objects

Throughout the nineteenth century, astronomers sought and discovered hundreds of asteroids between Mars and Jupiter. A few were even farther afield, as we now know, and crossed inward of Mars or outward past Jupiter. Some which were found followed the same general orbital path of Jupiter but were spaced at points either 60 degrees ahead of Jupiter or 60 degrees behind the giant planet. To distinguish these special objects, they received names out of Homer's *Iliad*, names of Greek and Trojan warriors at the battle for Troy more than three thousand years ago. The two groups thus became known as *Trojan asteroids*, and the two points—L4 and L5—became known as *Trojan Points*. Two dozen are officially named, with most of the Greeks in front and most of the Trojans behind Jupiter—the Greeks are either leading, or running away from, the Trojans, depending on your sympathies in that ancient war.

About seventy Trojan asteroids have been charted, and a great many more have been spotted, down to the limits of resolution. Oddly, there are several times more objects in the leading group than in the trailing group, and no explanation has been generally accepted.

Astronomers have asked, If Jupiter has Trojans, why can't other planets? Calculations showed that the perturbing forces of the Sun and Jupiter itself would displace objects in Earth's or Mars's or Venus's L4 and L5 points; asteroids, dead comets, and miscellaneous solar system flotsam and jetsam might temporarily gather there—but they would be too small to be seen from Earth anyway. There has been speculation that the Earth-Moon L4/L5 points held natural objects, or even alien artificial objects; one experiment on Skylab scanned these zones and determined that there was nothing there over ten feet in diameter.

The Voyager probes to Saturn finally found other natural *Trojan objects*. Saturn, of course, possibly has some of its own, ahead of and behind in its thirty-year orbit. But the Trojans spotted by Voyager belonged to some of Saturn's own moons. Observations from Earth had spotted two such objects in 1980, just before the first fly by, but the spacecraft discovered at least three more, and the pictures are still being examined. Dione has at least one preceding Trojan object (at L4), and Tethys has at least one each in L4 *and* L5.

These discoveries confirmed that nature does speak the language of mathematics after all, and that Lagrange was a good interpreter!

The Planet Clarion

Part of the modern folklore of space involves a planet called Clarion located at the Sun-Earth L3 point; that is, it is on the far side of the Sun as viewed from Earth, and hence would be unobservable from Earth. Several UFO crackpots claim to have been visited by spaceships from the phantom planet, and renowned psychic Jean Dixon has pinpointed that location as the home base of flying saucers (which, Dixon promised in 1976, were "piloted by women" who would unveil themselves to the world in 1977—a kind of cosmic striptease that strangely never happened). The planet was even the theme of a 1960's science fiction movie, *Journey to the Far Side of the Sun*.

But Lagrange knew better, and so do we. Since L3 is not a stable point, perturbations on the object due to the gravity of other planets would pull it out of position within a year and would allow it to be spotted during total solar eclipses. More telling, its own gravity would have noticeable effects on other planetary orbits within a matter of months. Sorry, Clarion—L3 is empty, at least of natural objects.

USING THE LAGRANGIAN POINTS

When space scientists wanted to "hang" a solar storm warning probe between Earth and the Sun, they chose a modified Lagrangian trajectory. In 1978, the International Sun-Earth Explorer 3 (ISEE-3) was launched on a long looping trajectory which took it, a few weeks after blast-off, through the Sun-Earth L1 point. There, maneuvering rockets were fired to allow the probe to dwell in the region. The exact L1 point was not feasible since it would be viewed from Earth directly into the Sun, thus drowning out the probe's radio signals with solar static. Instead, it was placed in a "halo" orbit in which it circled the L1 point as viewed from Earth. Positive control from small jets was needed to keep it from drifting away. In 1983, its solar mission completed, the probe was steered past the moon, and it took off toward the comet Giacobini-Zinner.

Such halo orbits have been proposed for lunar far-side communications relay satellites. That would be at the Earth-Moon L2 point, which would for all practical purposes be blocked from sight from Earth. Instead, the probe would circle the L2 point using control gas for five or ten years before being replaced; it would provide communications with far-side vehicles, including both landers and orbiters (real-time signals from orbiters, relayed through another satellite, are the only way to chart far-side mascons and other gravitational irregularities fully.

Recently, Soviet space engineers have tagged the Earth-Moon L1 point as a feasible staging area for a manned lunar landing expedition. During Apollo, the staging area was a spacecraft in low circular orbit above the Moon; this allowed the descent vehicle, the Lunar Module, to be as light as possible, but it introduced navigational problems and severely limited the regions which could easily be explored. Although a mother ship at L1 would entail a beefed-up landing craft, it would allow a greater variety of landing sites and a practically continuous access to the surface and back, without restrictive and narrow launch windows.

Soviet mathematicians, well aware that L1 is an unstable point, have calculated that over a year's period, keeping a vehicle within ten or twenty miles of the L1 point would only require 20 to 30 feet per second of *delta-V* (velocity change needed to alter its course), or an

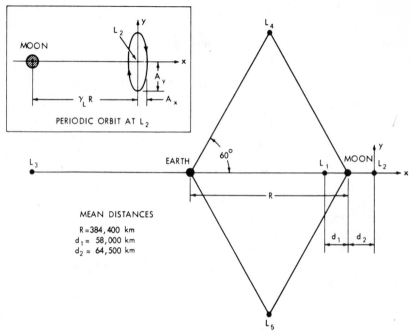

The Lagrangian Points in the Earth-Moon system. They apply equally to any two bodies, and a possible Trojan asteroid for Mars was reported early in 1984. (Illustration courtesy of NASA.)

amount of propellant equivalent to less than half of one percent of the spacecraft's total mass–an entirely reasonable operating cost.

During an internal power squabble in the L5 Society (a space colonization enthusiast group), one dissident faction set up an alternate communications center for coordination of national activities. Whether is was "stable" or "unstable," it showed a sense of humor that would have made Lagrange chuckle at how far his mathematical elegance had gone. The group's name: Elf Hive. And that's the last word in Trojan Points!

17

Spaceships of the Future

"A ten-month trip to Mars is nuts," snorts Robert Bussard. Instead, he proposes a turn-of-the-century spaceship which could make the trip in three weeks. "This is the way to do it," he continues, "with point-to-point travel times equivalent to those of the old West."

Despite today's temporary retreat in far-sighted national space commitments, there is no lack of ideas for future interplanetary spaceship propulsion. They range from Franklin Chang's "ion drive with an afterburner" to Rod Hyde's laser-pulsed fusion drive (a remote but lineal descendant of the original Orion drive design of the 1950s), to Robert Forward's multidimensional fascination with antimatter annihilation drive. They all might work, and others with them, when the time comes for really serious interplanetary trucking. It's safe to suppose that at least some of them will work, and then the Solar System will take on human dimensions.

Duration, not distance, has long been the true measure of geographical separation on Earth. Years-long reconnaissance voyages around the world can now be made at leisure in a few months—and hurriedly in forty hours. Transatlantic excursions can now be a matter of weekends, not months.

In the spring of 1983, Pioneer XI departed the Solar System after coasting up and out for eleven years. Had the spacecraft been able to maintain the high acceleration which characterized its first ten minutes of flight, the trip would have taken about a week.

HIGH-EFFICIENCY ENGINES

The time will come when voyages to Mars no longer consume the better part of a year. With the right kinds of engines, people could get from here to there and back again within the span of a summer vacation.

As discussed in chapter 2, some simple calculations can demonstrate the true point-to-point durational dimensions of the Solar System if engine power were no longer a limiting factor. Since human beings will be aboard future spaceships, the ship's acceleration should probably be limited to the force of gravity on Earth's surface, or one-G. The flight plan would call for continuous thrusting, building up speed to the halfway point, then flipping around and braking to the destination.

With such an engine, a manned spaceship could reach the Moon in three hours and forty minutes instead of the three days needed by Apollo astronauts. With such an engine, a manned spaceship could reach Mars in fifty-five hours instead of the ten months it took the Viking robots. And with such an engine, a manned spaceship could reach the edge of the Solar System in sixteen days instead of the decade needed by Pioneer. With such an engine, even the nearby stars come within planning range.

That may be asking for the impossible, at least for the next century. But even an engine capable of a mere 0.01-G constant push would be nothing to sneeze at. It could get a spaceship to the Moon in thirty-five hours and to Mars in three weeks—remarkably rapid trips in terms of today's space technology. And such engines are almost certain to become feasible within a few decades.

Blueprints for such engines are now being drawn up. Their power plants are described in technical phrases such as *fusion pulse*, *annihilation drive*, and other esoteric terms, but they share one purpose: get from here to there and back as quickly and cheaply as possible—and before the next century is half over.

Spaceship engine power, like any other quantity in the statistical universe, can be measured and quantified. Numbers can be applied to it, and equations can simulate its behavior and effects. As mentioned in chapter 2, *specific impulse*—the duration, in seconds, of how long one pound of fuel can provide one pound of thrust—of a spaceship engine. Oddly enough, the unit of measurement of this quantity—abbreviated Isp and pronounced "eye-ess-pee"—is seconds of time.

In more precise engineering terms, Isp is equal to exhaust velocity of burned fuel from the rocket nozzle, divided by the acceleration of gravity (small "g") and by the molecular weight of the expelled material. Furthermore, combustion engineers know that exhaust velocity is proportional to the square root of chamber temperature. The strategy should be clear: to increase engine efficiency, increase the exhaust velocity, increase chamber temperature, and use as light an exhaust material as possible—preferably, hydrogen gas.

Standard liquid-fuelled military rockets have an Isp of about 250 seconds, which they can maintain for several minutes. The space shuttle's main engines, burning liquid hydrogen, have an Isp of more than 450 seconds and fire for more than eight minutes. Nuclear rockets using fission to heat hydrogen were able to deliver 800 to 1,000 seconds of Isp in ground tests in the late 1960s, in burns lasting tens of minutes. *Ion drive*, or *electric propulstion*, offers an Isp of 10,000 seconds or more, for weeks or months on end, but with all the push of a butterfly's kiss.

The engine of the semimythical Solar System spaceship would need an Isp of one million seconds. The question is, How do we get there from here? There are a great many candidate answers, only a few of which will turn out to be right.

The Hybrid Plume Plasma Engine

To demonstrate a confidence in a proposed new rocket propulsion scheme, perhaps a specialist should be willing to ride on it. And there is indeed one such propulsion researcher who also happens to be an astronaut: Dr. Franklin Ramon Chang, a Costa Rican–born physicist now a member of the NASA astronaut corps in Houston.

Scientist-astronauts (now called Mission Specialists) have always been encouraged by NASA to pursue their outside scientific research along with their flight training. For Chang, who was picked as an astronaut in 1980 and is now working on the research program for a 1985 Spacelab mission, that research has primarily been a space propulsion idea he first developed while working with MIT researchers.

"Basically, it's an ion drive with an afterburner," explains Chang. Called a Hybrid Plume Plasma Engine, the rocket takes an ordinary high-temperature plasma generator and couples it to a nozzle system designed to inject inert gas around the plasma jet. The combined plume then pours out a rocket nozzle, generating much more thrust than the plasma jet would have alone.

"I consider the number one feature of this concept the variable specific impulse," Chang explained. "The second major feature, which is related to the first, is the non-uniform temperature gradient of the exhaust plume"—it's much cooler along the edge than it is in the center.

Neither concept is original, but their combination is. In addition, Chang has pioneered work in the computer modeling of the interaction of the plasma jet with the inert material surrounding it. Although *film cooling* has been seen before in rocket designs, it has not been over such extreme temperature ranges. The result is an engine which may be able to propel both interorbital "space tugs" as well as fast interplanetary probes, with significantly larger payloads than those possible with current systems.

"In a lot of this stuff, I'm still a Lone Ranger," Chang admitted. "We don't have all the numbers worked out yet." But with the help of a graduate student at MIT, and with serious interest from NASA's Jet Propulsion Laboratory and the Air Force Rocket Laboratory at Edwards AFB, Chang has seen the concept evolve from an idea to a plan for laboratory testing of the hybrid plume stability.

Such testing could come by 1985. Following that, Chang envisions an orbital demonstration with a small hybrid plume plasma engine attached to a platform in the space shuttle's payload bay—perhaps with astronaut Chang along to assist in the test.

"By varying the amount of inert material injected, the Isp and thrust can be set to optimal levels," he explained. High thrust at the beginning of the trip would be followed by low thrust for the long haul. Upon arriving at a gas giant planet, a vehicle with such an engine could scoop up atmospheric gases and use them as propellant for the return trip. "The engine will work with just about any gas, if tuned properly," Chang asserts.

The current limiting factors are associated with the engine's electrical power source which creates and expels the plasma jet. Chang foresees first the use of nuclear reactors, which under current state-of-the-art could allow the engine to operate with an Isp ranging from 500 seconds (not much better than chemical systems—but it can use almost anything for propellant, even gas snatched from the upper atmospheres of giant planets or Titan!) to 5,000 seconds for the really serious interplanetary trucking.

"This is not fusion power," Chang is quick to point out. "But if we knew how to do fusion power it would work even better." And when fusion drives are built, they may be advanced versions of his hybrid plume design. For that matter, Dr. Chang, who is still in his early 30s, may help build them—and ride them.

Alternative Propulsion Energy Sources

While science fiction spaceships tend to zoom through Hollywood skies in what may graphically be called the "belchfire mode," there are likely to be other technologies for interplanetary travel or for important segments of such travel. Some may indeed look like cosmic torches with their fiery exhausts, but others may not push but be pushed— by sunlight: the *light sail* concept. There is even one idea which involves not pushing at all, but pulling: the old *skyhooks*, renamed *space tethers* (see Chapter 18).

So as not to miss a trick with potential non-obvious space propulsion breakthroughs of the next century, the Air Force's Rocket Lab recently funded a series of study contracts. One of the recipients of such a grant was Dr. Robert Forward of the Hughes Research Labs in Malibu, California. Dr. Forward promptly took a year's leave of absence to survey the entire field of futuristic concepts.

The study is called "Alternate Propulsion Energy Sources," and the charter was to examine any new idea for extracting, storing, or using energy in space.

"I collected about sixty ideas," Forward told me, "of which I looked carefully at twenty-six. And from them I chose four as deserving of serious additional investigation.

"I looked at the physics side of the question, not the mission side," Forward explained. "I wasn't thinking applications at all—there was no conceptual baseline mission."

The four selected ideas, which reflect the wide spectrum of possibilities are (1) solar-pumped plasmas for efficient electrical power generation, (2) perforated light sails, (3) a metastable fuel still under a proprietary study, and (4) Forward's old favorite, antimatter propulsion.

The first idea involved a much more efficient method of converting sunlight into electricity than solar cells, which are now used on most satellites and space probes. The electrical power would then be used with other forms of electrical plasma propulsion, such as Chang's drive. By itself it is not a propulsion scheme, but it promises to make other schemes needing large power sources much more efficient.

The idea for light sails full of holes was an intriguing one. "I invented it," Forward pointed out, "although Freeman Dyson thought of it too." Light sails have already been shown to possess highly attractive super-long-haul cargo-carrying ability, even without such improvement. Perforating a large *solar sail* (Forward prefers the term light sail since "I intend to push it with other energy sources, such as laser beams") could reduce its mass by 90 percent while not reducing the sunlight pressure—as long as the holes were smaller than the wavelength of light.

"The holes might provide another advantage," suggested Forward. In Earth orbits below 1,000 kilometers, light sails cannot be deployed because of the retarding effect of collisions with air molecules. But a perforated sail might not be nearly as susceptible to interference; Forward suggests a space shuttle experiment as soon as possible, to hold up a sample of such a sail and measure its actual sunlight push and air drag.

The third idea—metastable fuel—is not ripe for discussion due to ongoing research, but the concept of *metastable states* of matter is not new. The energy required to convert ordinary substances into such material could subsequently be released as thrust—if it could be safely bottled up until that moment. Forward's report described what is essentially "solid fuel" (to use it, says Forward, "you burn it") with an Isp of several thousand seconds! It could be used for spaceships climbing into orbit from gravity fields as strong as Earth's or stronger.

Fusion Energy

There are, of course, other propulsion possibilities. Fusion energy is the most obvious. "It's also worth looking at," Forward suggested, "once it gets going—but it hasn't yet. It is definitely a very viable technique."

Rod Hyde certainly thinks so. Hyde, a physicist at the Lawrence Livermore Laboratories in California, is working on laser-induced fusion for commercial power generation and is also interested in future spaceship propulsion designs. He has helped conduct several advanced studies along those lines. In October 1983, he presented a report to the International Astronautical Federation annual convention in Budapest.

"The key thing is that most of the development costs are being paid by other people," Hyde pointed out. "Once laser fusion works on Earth, it becomes trivial to build a spaceship."

The interplanetary spaceship Hyde envisions (based on studies begun in 1980) is a thousand-ton space freighter that can haul cargo throughout the Solar System at 0.1-G acceleration. "That's two weeks to Mars, five weeks to Jupiter, and twenty-five weeks to Pluto," noted

Some of the most promising advanced propulsion systems must be built large in the beginning. That could take years—but then the Solar System liners could make repeated roundtrips to Mars within weeks. (Painting by Denise Watt-Geiger, courtesy of NASA.)

Hyde. "The trajectories aren't very sexy—almost just straight lines." But the payload fraction very definitely is sexy—about 50 percent, by Hyde's calculations.

Laser-induced pulse fusion is the key. Fittingly enough, that concept itself got its impetus from the seminal Orion concept for a spaceship propelled by exploding thermonuclear bombs. In the version that may well fly, the explosions are much smaller and much more frequent ("tens per second" is all Hyde can say), and they are to be caused not by detonating bombs but by the ignition of tiny pellets by aimed lasers. The desired result is obtained: mass is expelled at very high velocities, and appropriate mixtures of pellet material can produce appreciable thrust levels.

"There is one main technological difference between laser fusion for power and laser fusion for space propulsion," explained Hyde. "In space, you must reject the waste heat, so the laser must operate at high temperatures." This is satisfied by certain types of lasers called rare gas excimer lasers—but they are not among the current contenders for power generation. Work is progressing on them for possible military applications, and Hyde believes they will become attractive for operational fusion power plants whenever such systems are designed.

The next step is to adapt the power plant for space. "There is no market for such a rocket now," understates Hyde. But the availability of such a rocket, if widely known, can contribute to a demand for one in coming decades.

According to Arthur C. Clarke, prophets of technological progress tend to be overoptimistic in the short run and underoptimistic in the long run. One of the reasons for the latter failure is the existence of people like Robert Bussard. He's also talking about revolutionizing Solar System transportation within the next human generation.

Currently Bussard is working on what he calls a Riggatron Fusion Tokamak, a copyrighted name for a revolutionary new fusion power generator. "It's a small high-field copper-coil tokamak device," he explained recently, "running on deuterium-tritium fusion. It makes cheap steam and cheap neutrons, and produces energy at a cost equivalent to a two or three dollar a barrel of oil price."

By the end of the 1980s, Bussard expects to be working with experimental units, and within four or five years after that, the first commercial power plants should be available. "If that works," he continued, "we can push it to work on deuterium-deuterium reactions alone, with higher efficiencies."

Bussard went on: "If the coils are then extended, and we let plasma leak out past magnetic diverters, and mix it with hydrogen, we have a spaceship drive." Various combinations of Isp and thrust-to-weight ratios are possible, he asserted, with the peak efficiency at about 6,000 to 7,000 seconds and a thrust of about 20 percent of the engine's weight.

"For a spaceship, that's ten times the force of the Sun's gravity out here by Earth, so it's a 'high thrust' system even though the acceleration is only several milligees"—about a tenth of a foot per second per second. "Because of the minimum size of the fusion device, we would have to build a five thousand–ton spaceship," Bussard went on, "but it could get to Mars in twenty days with a payload fraction of one quarter." Bussard's group figures that the transportation cost from low-Earth orbit to low-Mars orbit would be only a few dollars per pound.

At Mars, the spaceship could use its power plant to generate electricity and electrolyze water (found inside the moonlet Phobos, or in subsurface permafrost). This would create all the hydrogen propellant needed for the homeward leg.

How soon could this be done? Bussard is optimistic: "We'll have deuterium-deuterium reactions running in the mid-1990s on the ground. Within ten years it could be turned into an engine. If anybody wanted to build a spaceship, in twenty to twenty-five years from now we could have such a machine."

The advantage of fusion, Bussard pointed out, was in the energy bonus. Any electrical propulsion system uses electricity to expel propellant, with varying levels of efficiency. The Riggatron Drive would use electricity to drive a fusion reaction. That reaction in turn makes more energy which uses thermal processes to push out the propellant— but the fusion reaction introduces a factor of twenty or more gain in the original energy input. Concludes Bussard, "We see it as rationally possible with the physics we know and with the engineering we know. It's just time and money."

Or maybe not. This conflict between Bussard's continuous magnetic fusion and Hyde's pulsed fusion is not easy to resolve, but Hyde seems to have access to incredibly powerful and high-fidelity computer programs.

"The weight of the shielding for the super-conducting magnets is one objection," Hyde complained. That would affect the power-to-weight ratio of the system as well as the amount of absorbed waste heat which would need to be rejected.

Nor is tritium such a readily available fuel. To avoid the need for prohibitive amounts of it, Hyde devised a scheme in which neutron-bombarded lithium produced sufficient tritium gas for "salting" the fusion pellets prior to their insertion into the thrust chamber. The spaceship would have to be a flying fuel-processing plant as well—but such an idea might vastly improve its financial attractiveness.

Antimatter

The most powerful spaceship fuel currently imaginable is antimatter. Reactions release up to a thousand times more energy per unit of fuel mass than a typical nuclear fusion reaction. As a spaceship fuel, antimatter does have drawbacks: it must be manufactured by techniques currently unknown; it must be stored safely by techniques currently unknown; and it must be transformed into propulsive force by techniques—you guessed it—currently unknown.

But such techniques are no longer unimaginable. Speaking to spaceflight enthusiasts of the British Interplanetary Society in November 1982, former NASA deputy administrator Dr. George Mueller asserted (based mostly on Robert Forward's work) that recent developments in nuclear physics had made such technology marginally conceivable at last: "I am not predicting that an antimatter space drive will propel us to Alpha Centauri in the next twenty years," he admitted. "But I am saying that the science of antimatter has advanced to the stage that it merits serious consideration indeed.

"The technology of producing antimatter with particle beams is under development at numerous high-energy particle laboratories around the world," Mueller pointed out, referring in particular to the European Center for Nuclear Research (CERN) near Geneva, and to America's Fermilab. Antiparticles can be created, "cooled," and then stored in magnetic confinement. So far, only the tiniest amount of such material has been handled.

The best spaceship propulsion technique would be to mix small amounts of antimatter with much larger amounts of an inert propellant, such as liquid hydrogen. "This arrangement should produce specific impulses in the range of several hundred thousand to several million seconds," noted Mueller. Therefore, "short duration Solar System voyages can be accomplished readily if antimatter can eventually be manufactured and handled in quantities of a few hundred kilograms." And that's the catch.

Research into fusion power and into beam weapons has importance for antimatter research, too. Ultimately, the artificial production of antimatter will require dedicated particle accelerators of very high currents and voltages. But the electrical power requirements alone are staggering.

Members of the British Interplanetary Society, who produced the Daedalus starship design in the mid-1970s, are not staggered by these problems; they are intrigued by them. One issue of their journal was entirely devoted to antimatter propulsion (September 1982), and it contained such papers as "Concepts for the Design of an Antimatter Annihilation Rocket," "The Cryogenic Confinement of Antiprotons for Space Propulsion Systems," and "Design Considerations for Relativistic Antimatter Rockets." The papers were highly technical and almost excruciating in mathematical detail—but they were well-founded speculation, not science fiction.

One design for an antimatter factory in deep space uses solar collectors 186 miles on a side to produce a power flux one hundred times the current power output of the entire world. At an efficiency of 0.1 percent, a high-energy proton beam would be used to manufacture one kilogram of antimatter every month.

"The sun pours 1.3 grams [0.046 ounces] of raw energy per day through every square kilometer [0.3861 square miles] of space," noted

Robert Forward recently. "If we can convert even a small part of that energy into antimatter, then we become lords of the Solar System. If we can do better, then we become tourists to the star worlds."

Mueller concurs: "Because of its unmatched capacity to carry available energy in concentrated form, antimatter should be expected to figure prominently in space propulsion for the twenty-first century. . . . Even modest progress with antimatter propulsion would revolutionize travel across the Solar System."

Such revolutions are doubtlessly coming. With such engines, even the mind-numbing millions of miles between planets can be overcome, and the Solar System—connected by travel times of weeks or months— can assume human dimensions. Once people grasp the concept of distance as a function of time and not space, a new age of exploration and settlement can unleash its tumultuous forces. That attitude, however, awaits the enabling technology of interplanetary engines. They, in turn, are not likely to wait for very long.

18

Tethered Space Operations

A very ordinary-looking hot air balloon floated peacefully between the dark blue sky and the thick yellow-tinged cloudbank below. Underneath the sphere a small gondola dangled motionless.

Preceded by a sharp sonic boom, a small arrow-shaped vehicle appeared in the sky, plunging nearly straight downward. Trailing behind it was a nearly invisible filament, a line connecting the arrowhead to something else on the other side of the blue sky.

The arrowhead's path angled directly toward the peacefully floating balloon. Its descent rate slackened due to increasing tension on the line trailing behind it. Guided by small fins steered by an onboard computer once designed for precision reentries of thermonuclear warheads, the probe homed in on the balloon and pierced it amidships.

As the balloon collapsed, hooks deployed from the slowing arrowhead below it. The arrowhead—now a grapple iron—reversed its motion and began moving upward under the force of the line attached to the object above the sky. The hooks caught onto the skin of the balloon and pulled it, along with the still-attached gondola, higher and higher.

The skies portrayed in this vignette are those of Venus. In orbit around it, dangling a hook on a two-hundred-mile cable, is an unmanned spacecraft of the early twenty-first century. In the gondola are samples of the surface and subsurface rock of hellish Venus, snatched from the extremes of temperature and pressure by human ingenuity.

Standard scenarios of future spaceflight generally consist of bigger, faster, and more efficient spaceships, extensions of the technologies already familiar today. Despite the well-known historical observation that no other exploration and transportation evolution has ever occurred without unpredictable quantum jumps in technology, all too many spaceflight prophets fail to take into account the likely advent and revolutionary effect of completely new ideas.

One such idea, only slowly being recognized, may revolutionize space station architecture, earth-space-earth transportation techniques, space propulsion and power systems, and—as in this imaginative vignette—interplanetary probes. This is the idea of *space tethers*.

As applied to space missions, the term tether simply means a long line connecting two or more space vehicles. The line can be tens or hundreds of miles in length. Current materials—such as Kevlar—already can be applied to building tethers with potentially revolutionary applications to space operations, in the very near future.

Tethered space operations have already occurred. In 1966, two Gemini missions conducted experiments with short tethers to Agena target satellites. One, *Gemini 11*, performed a stabilization test where the Agena dangled underneath the Gemini, utilizing the gradient of Earth's gravity field to maintain position. The second test, on *Gemini 12*, involved slowly spinning the two multi-ton objects to generate a slight "pseudo-gravity" force.

THE TETHERED SATELLITE SYSTEM

A new generation of space tether activities is about to begin. In 1982, NASA and the Italian Space Agency signed an agreement to fly a joint mission in 1987 involving an Italian spacecraft and an American tether system. The spacecraft is to be extended up to sixty miles, both upward to investigate ionospheric physics and downward to make long-term measurements of the upper atmosphere.

The *TSS*, or *tethered satellite system*, consists of the deployed payload, a deployment boom extending upward out of the space shuttle's payload bay, and a reel assembly with tens of kilometers of cable. During the initial deployment (and during the final minutes of retrieval), small jets on the payload push it away from the shuttle, keeping the line taut. Once the objects are a few kilometers apart, differential gravitational effects will keep tension on the tether without need for jet firings.

Due to a balance of forces acting on it, the tethered object will hang nearly directly below (or above, if deployed in that direction) the space shuttle. The tether is not supporting the subsatellite's entire weight, and in fact is providing only a very small fraction of the force keeping the subsatellite up. Most of the "force" is due to the observation that the subsatellite is "almost" in orbit—that is, it is going only a little bit slower than a free-flying satellite would be going at that same altitude. The tension on the tether provides only enough force to make up for this small energy deficit.

That is an important point because it contains the seeds for an amazing series of new ideas using tethers for space operations. Here it is worded slightly differently: two tethered objects may be in a stable orbit, but the lower object is flying slightly more slowly than would a stable free-flying satellite, and the upper object is flying slightly more rapidly. The center of mass of the tethered system, located somewhere along the cable between them, is of course flying at precisely the correct velocity for a satellite at that altitude, whether it's a point mass or an extended structure.

Now imagine the tether is cut. Since the lower satellite suddenly becomes a free-flying object without sufficient speed to maintain its orbit, it falls into an orbit closer to Earth. Because the upper object has an excess of energy over that needed merely to maintain a circular orbit at its altitude, its orbit rises somewhat. These changes, of course, occur at the orbit points opposite to where the tether is cut; the point of cutting still remains on the objects' new trajectories.

As a rule of thumb, the separation of the two objects at the far side of their new orbits will be about seven times as great as their initial separation when tethered. If they were of equal mass and had been connected by a hundred-mile long tether, and had been orbiting at an average altitude of four hundred miles, the top object would rise from 450 miles (where it started) to 800 miles, while the bottom object would drop from 350 miles in altitude to zero—that is, it would fall back into Earth's atmosphere.

A modern space shuttle burns substantial amounts (about a third of its fuel tanks) of rocket propellants to perform its "deorbit burn" at the end of its mission. A decade from now, an American space station will need to carry a significant amount of rocket propellant to boost its altitude against the slow orbital decay caused by aerodynamic drag. But these two propulsive activites are in exactly opposite directions. One raises the orbit of an object and the other lowers it. If a space station were tethered to a space shuttle, then reeled the shuttle out downward to a substantial distance, and merely unlatched the tether at the end of the mission, the shuttle would deorbit and the space station would be boosted *for free*. Neither of the loads of propellant— the one for deorbit and the one for altitude raise—would be needed.

This is not magic or mumbo jumbo, it is authentic, albeit nonintuitive, physics. What is happening is that momentum of the entire system is conserved, but the tether allows some momentum to be transferred from the space shuttle (which then falls out of orbit without needing rocket fuel) to the space station (which then receives a routine orbital boost without using any fuel either). There is a cost, of course: the weight and complexity of the tether system. But recent studies have clearly demonstrated the cost-effectiveness of this strategy, a strategy which was practically unknown even a few years ago.

The inventor of the tether concept, along with almost every other basic concept in spaceflight operations, was the turn-of-the-century visionary Russian, Konstantin Tsiolkovskiy. But the champion of actually doing it has been Giuseppi Colombo, an Italian space engineer who played a key role in the initiation and approval of the tethered satellite program. According to Ivan Bekey, director of advanced programs and plans in NASA's Office of Space Flight, Colombo (who died early in 1984) "is truly the father of space tethers."

Colombo's Proposal

One example of Colombo's inspired engineering genius is a recent modest proposal for "A New Concept in Space Station Architecture." The basic idea is to connect sets of orbiting shuttle fuel tanks with a number of cables tens of miles long. Colombo and coauthor Jack Slowey of the Smithsonian Institution Astrophysical Observatory in Cambridge, Massachusetts, have issued a series of detailed reports on this concept. Although much of the reaction in the space community was very favorable, there were a number of discouraging criticisms. "The main difficulty," noted Colombo delicately, "is related to the fact that most people are not familiar with the principles of celestial mechanics"—and conclude erroneously that Colombo's concepts involve black magic, not real technology.

The complete structure proposed by Colombo and Slowey would consist of two sets of twenty-five tanks, each weighing thirty tons. Between the sets of tanks, arranged in a platform structure, are six

Kevlar cables each twelve miles long. The cables need to be only one-fifth of an inch thick since they are *not* suspending the entire bottom platform's weight. In fact, since acceleration at the lower platform would only be 0.02-G, less than a ton of "weight" would be stretching the set of cables.

This space station design would allow many desired operations to proceed much more conveniently than with contemporary designs. The bottom platform would be used for docking the resupply flights from Earth, and in fact the shuttles could actually "land" on the platform and be held in place by the slight pseudo-gravity acceleration. The upper platform would be an observatory (far from the contamination and bumpy motion of the "dock") and a launch center. From there, payloads bound for deep space would be unreeled on tethers several hundreds of miles farther out into space and then released with a "bonus boost" of transferred momentum. Freight elevators would climb up and down the cables between the decks. For true zero gravity, a third mobile platform, attached to the cables between the two decks, would position itself at precisely the center of mass of the whole complex.

The assembly of such a station could occur in a relatively short time, according to a scenario drawn up by Colombo. Every two months or so, routine shuttle flights on other missions would drop off their external tanks in a parking orbit. The first such tank would have a special reel/tether/grapple unit installed on it, with which it would unreel itself a hundred miles above the shuttle before detaching. The second flight would approach the first tank, which would unreel the grapple, attach it to the second shuttle, and pull itself back down to the shuttle. There, the second tank would be attached to the first tank, and then both tanks would unreel themselves back up again, lowering the shuttle towards deorbit at the same time. This maneuver would be repeated every few months for two years until a large depot of tanks was assembled. At that point, a manned habitation module would be brought up and a construction crew could rearrange the tanks into the proper configuration. Only two or three dedicated launches would be needed for this process; the other launchings, numbering a dozen or more, would have carried ordinary payloads and only incidentally would have dropped off their empty tanks during their missions.

Perhaps this concept is just a little too revolutionary for the first American manned space station, still in the design stage. But even now, serious planners are suggesting that the space station have detached modules connected by long tethers. So the idea has registered, however subtly, on the consciousness of the spaceflight industry.

Tether-related Projects

One other sign that the tether concept has become scientifically acceptable is that NASA is now holding conferences about it. In June 1983, a group of space engineers met at Williamsburg, Virginia, to conduct a workshop on "The Applications of Tethers in Space." One topic was using tethers for transportation functions in space; another topic dealt with what were called *constellations*—an old term applied to the new idea of many large structures in space connected by long tethers.

The first topic was the delivery of shuttle payloads to higher orbits: The shuttle would reel out the payload on a 150-mile tether and unhook it at the far end of the tether, thus flinging it much farther into space proportionately to the relative mass of the shuttle and the payload. The shuttle would then reel the tether back in and return to Earth, saving propulsion in both the payload (perhaps a communications satellite on the way to geosynchronous orbit) and the shuttle itself. Conferees judged this technology to have "very high" practicality.

Another proposal called for tethering the space shuttle to its external propellant tank, which after blast-off is normally dropped off short of stable orbit. By carrying the tank directly into orbit on an optimized ascent trajectory, the shuttle could carry more payload as well. Subsequently, the tank would be dangled on a 25-mile tether, which would then be cut at a point appropriate for the tank to drop into the atmosphere over an empty region of the ocean. The shuttle and its payload would pop up into a higher orbit at this point, thus saving additional propellant. This tactic was judged to have "medium practicality"—the problem of attaching the tank to the shuttle and the impact of the tank's mass on the shuttle's attitude control system were serious questions.

Using tethers between a space station and a visiting cargo flight by a space shuttle was also a significant topic, with "high to medium practicality." The supply shuttle need never even dock with the space station, but instead attach itself to a space dock hanging many tens of miles below the station proper. This would save energy (and hence increase payload) while definitely enhancing the safety of such operations.

Some ideas were even more original. Jerome Pearson, already known for some excellent work on skyhooks and other advanced space concepts, suggested lowering an aerodynamic sail from a space vehicle into the upper atmosphere. The sail, a hypersonic airfoil built to withstand the severe heating induced by its speed, would allow the space vehicle to gradually turn left or right in its orbit, changing plane or inclination. Such a technique was deemed of "high practicality" since plane change maneuvers are prohibitively expensive if done with rocket engines.

The newfound enthusiasm of the space engineering community for tethers is also reflected in the birth of a new family of space-age acronyms for tether-related projects. There are TETRA (Tether Applications to Transportation), ELIOT (Electrodynamic Interactions of Tethers), GUT (Gravity Utilization through Tethers), TEASE (Tether Application in Space Experiments), TESCON (Tethered Spacecraft Constellations), and many more.

Another interesting possibility arises if a long tether is made of an electrical conductor. It then becomes a wire moving through a magnetic field (Earth's), which makes it an electrical generator if a closed circuit can be established. As described by NASA futurist Ivan Bekey, "If electrons are collected by a metallized film balloon at the upper end and ejected at the lower end by an electron gun, a current will flow downward through the wire." For a sixty-mile cable, the current will be about 15,000 volts at 5 amps, producing a net power to the payload of 70 kilowatts, many times the power level of current space shuttles. But of course the energy is not free—it is extracted from the momentum of the orbiting object, which drops by six miles a day.

Since the momentum-to-electricity conversion is so efficient (about 70 percent), it could be supplemented with a low-thrust high-efficiency ion engine to boost the station's orbit continuously while the orbit draws electrical power.

The reverse is also possible. With a high enough power source (solar cells, or a nuclear reactor), electricity could be pumped into the cable resulting in a net acceleration. The tether thus becomes a giant electric motor in orbit, using Earth's own magnetic field as the armature!

For short periods of time, this could provide emergency power to a vehicle whose main power system had failed. A simple wire could be extended and altitude could be exchanged for kilowatts for several days while repairs or rescues were being effected.

All of these concepts pale to insignificance next to the notion of dropping a tether from a satellite all the way to Earth's surface. If the anchor were in a synchronous 24-hour orbit, 24,000 miles above the equator, the cable could be anchored to a spot on the surface—perhaps Mount Kilimanjaro, or somewhere in Sri Lanka (a la Arthur Clarke's novel, *Fountains of Paradise*), or in the city limits of Quito or Singapore.

Since the cable would have to support its own weight (it would get no help from orbital velocity—it would be stationary), it would have to be tapered, thinner at the bottom and thickest at the top. And the required material strength is far beyond anything yet available.

But the concept of such a skyhook has been around for decades. Yuri Artsutanov first mentioned it in print in a Soviet magazine in 1960, and it has subsequently been independently reinvented several times. It's also been called a "space elevator" (transportation would be by "funicular railway" up the line) or, more in keeping with the semi-legendary nature of the structure, a "beanstalk." And, for the foreseeable future, it cannot be built on Earth.

However, there are other variations of the idea. Amazingly, they are entirely feasible with modern materials such as Kevlar.

Many military pilots owe their lives to a simple skyhook device which plucked them out of enemy territory when their planes had been shot down. They released a helium balloon attached to a long cable, hooked at the other end to a harness around their bodies. A rescue aircraft with a pronged fork extending below and ahead of it made passes at the balloon, and once they caught hold of it, pulled the pilots to safety.

In the Arctic, bush pilots have perfected a trick for dropping off and picking up mail, supplies, and other cargoes to isolated settlements. The aircraft lets out a bucket on a long line and then begins flying a tight circle around the bucket, which falls to the ground. While it bounces around at the end of the cable, a person rushes over to it, grabs it, retrieves its contents, and loads any return cargo. The exchange complete, the aircraft flies off straight and reels in the bucket.

Similar operations are feasible with space-based tethers. Some are difficult on Earth, with its deep gravity well (the deepest of any world in the Solar System on whose surface humans will ever walk), but are extremely attractive for smaller worlds such as Mars or the Moon. In fact, a synchronous skyhook could be built over the Moon with Kevlar; no stronger material would be needed.

For bigger worlds, a "rolling skyhook" may be one answer. This concept, orginated by John McCarthy of Stanford and elaborated by Hans Moravec, involves a long tether orbiting the Earth (or any planet for that matter) while also tumbling end-over-end. The tumble is made in the plane of the object's orbit, and its length is about twice its altitude. That means that whenever the tether is aligned exactly vertical, one end is actually touching the surface of the planet. The tumble rate is also adjusted so that in such a situation the end of the tether is moving backward at the same speed at which the tether's center of gravity is orbiting the planet—so that the tip of the tether is motionless relative to a point on the surface (once the planet's rotation speed is also taken into account). From a spot on the surface, the tip of the tether will appear to descend from the sky nearly vertically, come to a stop, and then accelerate back up into the sky.

The forces on such a rolling tether (so named because its motion looks like that of a diameter on a wheel rolling across a surface) are straightforward: the planet's gravity, the centrifugal force of the tether's own rotation, and aerodynamic effects, if any. According to Moravec, the minimum combined forces occur with a tether about one-sixth as long as the planet's diameter.

For Earth, that's a tether more than one thousand miles long. Also, for optimal strength, the tether must be tapered with the thicker, stronger part in the center (to support all the weight of the lower tether as well as any payload weight).

The skyhook doesn't have to come all of the way to the surface, either. It could pick up an airborne vehicle in midflight, whip it up into space, and release it with a tremendous outward-bound velocity or return it to another point in the planet's atmosphere, almost half a world away and only a few minutes later. Or the skyhook could provide even more efficient operation (and be technically feasible on Earth much sooner) if it latched onto a small rocket ferry that went only halfway into orbit (such a rocket-skyhook combination has been proposed by Moravec and a colleague, Burke Carley).

At the other extreme is the tether concept illustrated by the fictional vignette which opened this chapter. An interplanetary probe, orbiting a target world, reels out a long tether with a weighted "sample tube" at the far end. Once spun up, the tethered sampler device (perhaps containing a small expendable guidance computer to guarantee a precise impact point) descends to the planet's surface, scoops up material, and is pulled back into space. The tether is partially reeled in, and a robot "spider" crawls down the line, recovers the sampler, and attaches a new one for another pass. This could be repeated dozens of times during a brief orbital reconnaissance mission.

The sampler's descent and ascent must occur rapidly, as the mother ship holding the end of the tether passes overhead. For target planets with thick atmospheres, this presents a serious problem because the atmosphere will slow the sampler or burn it up, well before it could reach the surface.

In such a case, a compromise must be reached. If the tethered sampler cannot reach the samples on the surface, they must be brought part way up to the end of the tether. For a thick-aired planet such as Venus or Titan (or for deep atmospheric samples from Jupiter), a balloon may be the answer. The samples are grabbed by a short-lived robot dropped untethered to the surface: they are then loaded into a balloon gondola which is quickly dispatched back up to the upper levels of the atmosphere. There, the gondola is snared by the descending "smart skyhook," and yanked out into space. For the return to Earth, the sample cannister and a guidance system installed by the tether spider can be flung free of the planet's gravity with little additional rocket thrusting.

For target worlds such as Venus, the technological challenge of recovering surface samples has been awesome and intimidating, as long as mere rocket propulsion was considered. But as space engineers are learning, there are other ways than rockets to move mass through the Solar System and to and from planetary surfaces. The tethered satellite of 1987 is just a foot in the door for a vast future of this breathtaking new idea in space transportation.

19

Space Pilgrim's Progress

All of the retrospective surveys of the seventies and the forecasting expectations of the eighties with which the news media busied itself in 1979–80 should have been considered in the proper perspective. For the most part, they concerned themselves with trivialities, only occasionally tripping over some trend or development which will have any significance at all in a generation or two.

To my knowledge, absolutely nobody even obliquely referred to what the seventies will be best known for a century from now: It was the decade during which humankind learned how to live in outer space. The conceptual revolution—from viewing our world as a closed planet to seeing it as a gateway to the universe—has occurred in the minds of most educated people . . . all because of the events of the last fifteen to twenty years.

Consider where we were at the beginning of the seventies, just after the first Moon landing. Cosmonauts and astronauts had spent the sixties learning how to survive in space just long enough to conduct brief sorties to the Moon or for rushed maneuvering near Earth. The longest flight had lasted less than half a month, under conditions of significant discomfort and minimal habitability. And a flock of fearful medical problems lay waiting like dragons along the path toward the dream of space stations and interplanetary flight.

Ten years later the whole view had changed. Men had stayed in space for up to half a *year* and had engaged in productive work while remaining comfortable, healthy, and surprisingly happy. The way was cleared for the permanent occupation of near-Earth space, which surely will occur before this present decade is finished—a prelude to interplanetary expeditions lasting for years at a time.

Thus the seventies saw the opening of the most significant new habitat for humanity since people moved out of caves—an epochal revolution apparently too significant for comprehension by most news writers, who concentrated instead on more mentally manageable topics, such as women's fashions, political scandals, sports, and miscellaneous wars.

It was a difficult road, the decade of the seventies. The Soviets started it, with the *Salyut 1* space station, and paid a terrible price in lives for reaching too far too soon. Over the following years they saw additional space-station setbacks, the extent of which they were fairly successful in hiding from public attention. In the United States the Skylab program showed that men were not only able to work in space but were also able to conduct crucial emergency repairs on their habitats. In the ensuing years the Soviets got their act together, picked up the American space-station baton, and broke new ground in space-station development. Meanwhile the United States was (and is) engaged in the difficult task of changing gears in space hardware, advancing to the revolutionary space shuttle generation.

So space is no longer a place to make hurried sorties for specific preprogrammed missions. It is now a place for marathon excursions, where humans have many months to slip into the right rhythm and where new equipment and experiments can be shuttled into space on supply ships.

The next stage is easy to predict: people—men and women—will move into space habitats on a semipermanent basis, just as people today move from one city to another for new jobs. These people will fill out their income tax forms, send their holiday greeting cards, and cast their absentee ballots from space. For them, space will become a place to live and to work, not merely to visit.

Gradually, some of these people (and the population of space on New Year's Eve of 1990 will probably number in two digits) will come to think of outer space as their preferred homes. They may extend their duty tours. Some may develop medical complications or suffer

Once they are aboard a space city, many workers may begin to think of themselves as space colonists. (Photograph courtesy of NASA.)

crippling accidents that would make the return to Earth hazardous or highly impractical.

So somehow, someday, some people, or an individual, will realize that they have no intention of returning to Earth. They will consider themselves to be inhabitants of space, whether by exile or by choice. They will be the first space colonists. Whoever they are, they are old enough to be reading these words, assuming they speak English—which may *not* be a valid assumption.

Now, these living conditions in space are not likely to be luxurious by American (or even Russian!) suburban standards although there are bound to be compensations. They will, however, probably be at least as livable as seventeenth-century Jamestown, eighteenth-century Novosibirsk, or nineteenth-century Salt Lake City. Modular space stations only a few miles above Earth's atmosphere may eventually (more quickly than we probably expect) evolve into distant, large, spinning, Tsiolkovsky-Cole-O'Neill "space colonies." The breakthroughs, however, will be primarily philosophical, not architectural: they will be homes, not hotel rooms. And somehow, somewhere, we are bound to see the ultimate symbol of humanity's commitment to off-world migration: the birth of the first space-baby, and then the second, and the next, and the next. Young people reading these words will be their parents.

That's what the seventies opened the door for—without many people yet taking notice. The decade of the eighties is widening that open door and pushing down the walls that are holding us back. The nineties will unleash a flood so powerful that its echoes will resound throughout our entire civilization (or whichever civilization it is that is wise enough and farsighted enough to choose to turn the key)—so that even the newspapers will notice it! Space is out there now for those who would reach for it . . . to possess as a new arena, a new playground, a new home. The space race has no finish line.

Index

Abrahamson, General James, 130
Agnew, Spiro, 35
Albatros (Soviet shuttle), 119, 121, 131
Aleksandrov, Aleksandr, 101–16 passim
Ames Research Center (NASA), 68–69, 71
Andropov, Yuri, 98–99, 152
Antimatter, 28, 191–93
Apollo program, 24, 27, 32, 115, 153, 172, 184
Apollo 7, 41
Apollo 8, 16, 42
Apollo 13, 21, 49, 106
Apollo 15, 35
Apollo 17, 161
Apollo-Soyuz, 132, 134–35, 141–44, 149
Asteroid mining, 14, 175–80
Astronauts. *See also names of astronauts*
    Defense Department, 38–39
    payload specialists, 8, 36–39
    versus robots, 31–36
Atlantis (Orbiter), 21, 51–52
Aviation Week, 103–4, 109–10

Baikonur. *See* Tyuratam
Beggs, James, 141, 143
Beregovoy, Georgiy, 120, 122, 152, 168
Berezovoy, Anatoliy, 73, 78–90 passim
Beta cells, 10
Bone demineralization, 71–74
Bradbury, Ray, 173, 174
Brandenstein, Daniel, 25
Bussard, Robert, 183, 190–91
Bykovskiy, Valeriy, 149

Cape Canaveral. *See* Kennedy Space Center
Carrying the Fire, 15
CFES (Continuous Flow Electrophoresis System), 8
Challenger (Orbiter), 21, 51–52, 113, 119
Chang, Franklin Ramon, 183, 185–87
Chibis, 82
Chretien, Jean-Loup, 95
Clarke, Arthur C., 154, 190, 200
Closed loop life support, 87, 170
Collagen, 10
Collins, Michael, 15
Colombo, Giuseppi, 197–98
Columbia (Orbiter), 21–23, 52, 53, 64, 113, 119, 128, 148
Coolidge, Calvin, 51

Cosmonauts
    as propagandists, 145–49, 152
    guest, 12, 135–36
Cowings, Patricia, 68
Crippen, Robert, 23, 148

Discovery (Orbiter), 21, 51–52

Ejection seats, 23, 24
Enterprise (Orbiter), 23, 119
Erythropoitin, 10
Escape tower, 24, 113–16

Feoktistov, Konstantin, 105–6, 114
Fisher, Drs. Anna and William, 67, 74
Flagg, Richard, 111
Flying saucers (over Russia), 46
FOBS (Fractional Orbit Bombardment System), 126
Forward, Robert, 187–88, 191–93
Friction, myth of atmospheric, 26
Frosch, Robert, 122, 171

Gagarin, Yuri, 46
Gardner, Dale, 71, 76
Gazenko, Oleg, 72, 168, 173
Gemini program, 24, 41, 195
Geography, space, 27–30
Geosynchronous orbits, 13, 48–50, 154–60
Giacobini-Zinner (comet), 181
Grechko, Georgiy, 145–46, 152
Gulag, 43

Halley's comet, 2
Hartsfield, Hank, 148
Henize, Karl, 171
Hyde, Rod, 183, 188–91

Ivanchenkov, Aleksandr, 105

Johnson Space Center (NASA) 22–24, 36–39, 40, 77, 156–57, 162, 173
Johnson & Johnson, 7–8
Joint Endeavor Agreement, 7

KAL-007, 152
Kapustin Yar, 47, 123, 124, 125
Kennedy Space Center (NASA), 41–42
Kerwin, Joseph, 65
"Killer satellites," 146, 148